SHIPS OF WOOD AND MEN OF IRON

SHIPS OF WOOD AND MEN OF IRON

A Norwegian-Canadian Saga of Exploration in the High Arctic

by Gerard Kenney

Canadian Plains Research Center 2004

Copyright © Gerard Kenney 2004

Copyright Notice

All rights reserved. No part of this work covered by the copyrights hereon may be reproduced or used in any form or by any means — graphic, electronic, or mechanical — without the prior written permission of the author. Any request for photocopying, recording, taping or placement in information storage and retrieval systems of any sort shall be directed in writing to the Canadian Reprography Collective. The author may be contacted care of the publisher:

Canadian Plains Research Center
University of Regina
Regina, Saskatchewan S4S 0A2

National Library of Canada Cataloguing in Publication

Kenney, Gerard I., 1931–
Ships of wood and men of iron : a Norwegian-Canadian saga of exploration in the high Arctic / Gerard I. Kenney.

(TBS ; 9)
Includes bibliographical references and index.
ISBN 0-88977-168-5

1. Sverdrup, Otto Neumann, 1854–1930. 2. Fram (Ship) 3. Arctic regions—Discovery and exploration—Norwegian. 4. Arctic regions—Discovery and exploration—Canadian. 5. Canada—Boundaries—Arctic regions. I. University of Regina. Canadian Plains Research Center. II. Title. III. Series.

FC3963.K45 2004   917.19'5042   C2003-907140-5

Cover design by Brian Danchuk Design, Regina
Printed and bound in Canada by Houghton Boston, Saskatoon

We acknowledge the financial support of the Government of Canada through the Book Publishing Industry Development Program (BPIDP) for our publishing activities.

DEDICATION

For my daughters, Amanda and Jessica

Photo Credits

Page 4: A picture of Otto Sverdrup, taken in mid-life. Source: Otto Sverdrup's *New Land*.

Page 8: The crew of Sverdrup's second Arctic expedition, prior to departure from Oslo, 1898. Source: Otto Sverdrup's *New Land*.

Page 13: Four of the expeditions dogs. Source: Otto Sverdrup's *New Land*.

Page 29. The *Fram* in Fram Haven. Source: Otto Sverdrup's *New Land*.

Page 30. Inuit visitors, spring 1899, Fram Haven. Source: Otto Sverdrup's *New Land*.

Page 31. Johan Svendsen's cross, Fram Haven. © Jerry Kobalenko.

Page 37. Ove Braskerud. Source: Otto Sverdrup's *New Land*.

Page 38. Ove Braskerud's cross in Harbour Fiord, 1899. Source: Otto Sverdrup's *New Land*.

Page 38. Ove Braskerud's cross, c. 1965. *Canadian Geographic*.

Page 42. On the trail with an odometer. Source: Otto Sverdrup's *New Land*.

Page 55. Isachsen and Hassel. Source: Otto Sverdrup's *New Land*.

Page 74. The *Fram*, frozen in ice, winter of 1901–02. Source: Otto Sverdrup's *New Land*.

Page 78. Starting off on a dogsled exploration trip, spring 1901. Source: Otto Sverdrup's *New Land*.

Page 95. The *Neptune*. National Archives of Canada.

Page 96. The *Neptune*'s crew in winter dress. National Archives of Canada.

Page 97. A.P. Low and officers, 1903–1904. National Archives of Canada.

Page 100. A.P. Low and crew at Beechy Island cenotaph. National Archives of Canada.

Page 106. The *Arctic*. National Archives of Canada.

Page 107. Inuit woman in traditional dress. National Archives of Canada.

Page 108. Inuit women at Fullerton Harbour, 1904. National Archives of Canada.

Page 108. Inuit of Fullerton Harbour, 1904. National Archives of Canada.

Page 109. Captain Bernier and crew at Winter Harbour. National Archives of Canada.

Page 112. Captain Joseph E. Bernier, 1923. National Archives of Canada.

Page 117. The *Karluk*. National Archives of Canada.

Page 118. Vilhjalmur Stefansson, 1914. National Archives of Canada.

Page 120. The *St. Roch* on patrol. National Archives of Canada.

Page 121. Henry Larsen. Canadian Parks Service/Environment Canada.

Page 124. Henry Larsen. National Archives of Canada.

Map insert (between pages 36 and 37): based upon material provided by Rafael Valenzuela of Ottawa.

Contents

Photo Credits vi
Foreword viii
Author's Preface ix
Introduction xi
Part I: The Norwegian Expeditions, 1893–1902
The Man and the Ship 3
Sverdrup's Second Arctic Expedition 7
The First Winter 17
The Second Winter 35
The Third Winter 59
The Fourth Winter 75
Home To Norway! 85
Part II: The Canadian Expeditions, 1903–1948
Canada Is Not Amused 91
Albert Low and the *Neptune*, 1903–1904 93
Joseph Elzear Bernier, 1904–1925 103
The Canadian Arctic Expedition, 1913–1918 115
The Twenty Years of the *St. Roch*, 1928–1948 119
Canada and Norway Negotiate 125
Epilogue 129
Endnotes 131
Bibliography 132
Index 133

FOREWORD

GERARD KENNEY'S BOOK IS A SUPERB ACCOUNT of courage and enterprise by a Norwegian expedition under Otto Sverdrup from 1898 to 1902. It reads like a novel in which it is difficult to know which is the more admirable: the courage and endurance of the crew of the *Fram* or the skill with which they did what the British navy had never done—adopted and applied the techniques for clothing, hunting and food of the Inuit, without which they could not have survived for so many months. It is a remarkable tale—the more important now because global climate change is affecting our polar regions more than any other areas of the earth.

Passages in the Archipelago, hitherto frozen and barred by nature to merchant vessels, may become in reality the "northwest passage" that was sought by the British navy for so long. If so, our claim to the Archipelago under international law may become subject to more challenge than during the years when the region has been of little economic interest to Canada and to any other country.

The United States has never accepted our claim that the waters within the Archipelago are internal waters of Canada within the outer limits of the Archipelago. That claim will become increasingly difficult to maintain if a still warmer global climate makes the right of transit passage attractive for merchant shipping between the far east and Europe. The Arctic route would be much shorter than any other.

While the claim of Norway that Kenney deals with was settled in an agreement in 1930 recognizing Canadian sovereignty over the Sverdrup Islands, the Norwegian government specified that its recognition in no way implied approval of the so-called sector principle. It was a warning then, and the "principle" has no status in international law now. This book is a reminder that Canada today must not fall into complacency with respect to sovereignty in her Arctic lands.

Kenney's account of actions taken by Canada after the shock of 1902 caused by the Norwegian discoveries and claims, will be new to almost every reader. It is hoped that *Ships of Wood and Men of Iron* will be read by many Canadians. Virtually all the information will be new to them.

Gordon Robertson
Deputy Minister of Northern Affairs and National Resources and
Commissioner of the Northwest Territories, 1953–1963

AUTHOR'S PREFACE

Over 100 years ago, in June 1898, Captain Otto Sverdrup and 15 crewmen put out to sea aboard the schooner *Fram* from the Norwegian city today known as Oslo. When they returned to Norway four years later, they came back with a record of geographic and scientific discovery, the richness of which is unparalleled in the annals of Arctic exploration. This book is the story of those four heroic years spent in the High Arctic and their impact on Canada's subsequent efforts to ensure Canadian sovereignty in the area of the Norwegian discoveries.

Maps have been included in *Ships of Wood and Men of Iron* to help the reader follow closely the exploration sorties made by Captain Otto Sverdrup and his men during the four years his ship *Fram* spent in what is now the Canadian High Arctic at the turn of the last century. An attempt has been made to include on these maps significant place names in the Canadian High Arctic mentioned in the text on Sverdrup's expedition. All quotations that are not referenced to another source are from Sverdrup's book in two volumes entitled *New Land*.

I wish to thank all those who helped one way or another in the making of this book. I am especially grateful to northern experts Graham and Diana Rowley whose help was invaluable in ensuring the accuracy of this book. Special thanks are due the Norwegian Embassy in Ottawa which, through the good offices of Their Excellencies, Ambassadors Jan E. Nyheim and Johan Lovald, provided financial assistance that enabled me to travel to Oslo and visit the *Fram* museum where the famous ship is kept, and to pace the very deck that Sverdrup and his men did more than 100 years ago in the frozen Arctic. Special thanks are due as well to today's Norwegian Ambassador to Canada, His Excellency Ingvard Havnen, and Vice Consul Ralph Nilson for facilitating the choice of a publisher. Kåre Berg, curator of the Fram museum, read my manuscript carefully and offered suggestions that contributed greatly to the historical and geographical correctness of *Ships of Wood and Men of Iron*. Arthur Andreassen of Ottawa, Ontario, and Wenche Linneboe of the Norwegian Embassy in Ottawa, were both particularly helpful in translating the documents that Otto Sverdrup deposited in Arctic cairns during his 1898–1902 expedition and which today reside in the National Archives of Canada. Doreen Larsen Riedel's help with the section on her father, Henry Larsen, was much appreciated.

Gerard Kenney
Ottawa, Canada
January 22, 2003

"...in my opinion, Otto Sverdrup was the most competent and practical of all the Norwegian explorers of that era, but being both shy and reticent, he was satisfied with taking a back seat and was of course overshadowed by such men as Nansen and Amundsen ... from my own personal knowledge of the Arctic, there is no doubt in my mind today, but that Sverdrup was the most versatile and competent of the three."

— from the late Henry Larsen's unpublished autobiography written in 1962

INTRODUCTION

IN 1880, THE BRITISH CROWN CEDED its remaining territories in North America to the recently formed Dominion of Canada. It did this because it feared the covetousness with which the United States of America eyed the Arctic lands. Canada, reasoned the British, was in a much better geographical position to watch over and protect the frozen northlands from foreign interests than was Great Britain itself.

When Canada inherited these northern tracts, she was not even aware of exactly what she had received. A large area of the High Arctic was still "terra incognita," a clean white spot on the map. No one knew the extent of the lands transferred to Canada for the simple reason that no modern Europeans had ever been there. In fact, approximately half the High Arctic as we know it today had yet to be discovered, let alone explored, by contemporary Europeans or North Americans of European ancestry.

For the next 22 years, from 1880 to 1902, Canada essentially ignored her newly acquired lands, doing nothing of substance to explore and discover them. Then in 1902, an event occurred that shocked Canada out of her complacency. Captain Otto Sverdrup of Norway and 13 crew members returned from a four-year voyage of exploration and discovery in the High Arctic aboard the now-famous wooden vessel *Fram*. During the trip, Sverdrup and his men discovered and explored some 260,000 square kilometres (100,400 square miles) of new Arctic land, including the western half of Ellesmere Island and three major islands—Axel Heiberg, Ellef Ringnes and Amund Ringnes—not previously known to the modern world. What really shocked Canada, however, was that Sverdrup claimed these islands for the Crown of his country according to the age-old principle of discovery and exploration, still in vogue at that time.

Prime Minister Sir Wilfrid Laurier and his government lost no time in instituting measures to counter the Norwegian claim and to reinforce that of Canada. Thus began an era of Canadian Arctic patrol and exploration in the closing years of a period when ships were still of wood and men of iron.

First came A.P. Low, followed by Captain Joseph Bernier, Vilhjalmur Stefansson and Henry Larsen of the RCMP. These men, and their crews, are true Canadian heroes, who unfortunately are no longer as well known by their countrymen as they should be.

In 1930, Canada came to an agreement with Norway that in effect settled Norway's claim to the High Arctic islands discovered by Sverdrup, finally confirming Canadian *de jure* sovereignty over the entire Arctic archipelago.

Ships of Wood and Men of Iron is the Arctic saga of Norwegian hero Captain Otto Sverdrup, his ship *Fram*, and her hardy Scandinavian seamen in the years from 1898 to 1902. It is also the story of the four Canadian heroes who followed them to ensure that the High Arctic regions remained Canadian.

PART I

THE NORWEGIAN EXPEDITIONS 1893–1902

– 1 –

THE MAN AND THE SHIP

JUNE 24, 1898, DAWNED GRAY AND DREARY in southern Norway when Captain Otto Sverdrup stood out to sea from the port of Kristiania—Oslo as it is known today—together with 15 crew members in the small and seemingly frail Arctic exploration vessel, the *Fram*.

The heroic age of Arctic exploration in sailing ships was quickly drawing to a close, but countries could still achieve glory and add to their geographical possessions by being the first to set foot on unexplored land. Glory and material gain were no doubt in the minds of Norwegian Consul Axel Heiberg and the Ringnes brothers of brewing fame—Ellef and Amund—when they sponsored the *Fram*'s trip. On the other hand, Captain Sverdrup's motivation in exploring the Arctic—this was not his first such trip—seemed more one of responding to personal challenge, like Mallory's motivation for his fatal attempt to conquer Everest—"because it is there."

The plan for Sverdrup's trip—his second northern expedition on board *Fram*—had nothing to do with exploring the Arctic islands north of Canada, at least not when he set out. In his own words:

> *the route ... was to take us into Baffin Bay, up Robeson Channels, and as far along the north coast of Greenland as possible before wintering. From there we were to make sled journeys to the northernmost part of Greenland and as far down the east coast as we could. There was no question of trying to reach the pole.*

Sverdrup's reference to the pole becomes clear when considered in light of his previous exploration trip—the first *Fram* northern expedition—completed

SOURCE: OTTO SVERDRUP'S *NEW LAND*
A picture of Otto Sverdrup, taken in mid-life.

just two years before. In the late 1800s, reaching the North Pole was still one of those plums that explorers dreamed about. He who would pick the plum would be assured of a place in the history books of the world and probably would not have to worry about money for a long time. The plum was so seductive and the race to pluck it so bitter that even today there is not absolute certainty about whether Peary or Cook reached it first. The prize was so coveted that one of the two lied about discovering it. In fact, it has been suggested that they both lied and that neither of them reached the pole.

The Norwegian explorer Fridtjof Nansen was one of those who got caught up in the race to reach the North Pole. Attaining the pole, however, had not been his objective when he set out for northern latitudes on the first *Fram* expedition in 1893, with Sverdrup at the helm, or so he said at the time. His stated objective had been to prove his theory on the Arctic Ocean currents. Extensive studies of the northern seas had made him believe that there was a current of ice flowing from Siberia north across the polar region, and then south along the coast of Greenland.

Nansen was supported in his beliefs by articles of clothing and other flotsam found on the southwest coast of Greenland. These objects were thought to originate from the ill-fated American expedition ship *Jeannette*, which had been crushed by the ice off the New Siberian Islands in 1881.

The Norwegian explorer devised a plan to test his theory. He would force a small ship into the ice pack north of Asia and allow it to drift with the ice and see where it ended up.

Nansen's plans included Sverdrup from the start, and together they designed the *Fram* and had her built especially for the task. The explorers set out in June of 1893. Nansen was in charge of the scientific objective of the expedition, while Sverdrup captained the ship. By September 1893, *Fram* was locked solidly in the ice pack north of Siberia.

After 18 months of drifting with the ice, the Arctic current had brought *Fram* to within some 672 kilometres (417 miles) of the pole, which was not as close as Nansen had hoped; and soon, the expedition would slowly start drifting away on a southbound current. But the pole was too close and the attraction too great. During the drift, Nansen had drawn up a daring plan to reach the top of the world. On March 14, 1895, he and a crew member, Frederik Johansen, left the ship and started a dog-sled dash for the pole, knowing full well that they would not be able to regain their ship, but would have to get home by their own means. Sverdrup stayed on board as captain of the *Fram*, and together with the remaining 10 crew members continued drifting across the polar basin.

For the next 17 months, neither Nansen nor Sverdrup knew anything of the other's fortunes as they continued on, each trying to eventually make his way back to Norway as best he could, one by dog-sled and kayak, the other by ship.

Nansen and Johansen whipped their dogs across the rugged ice to within 385 kilometres (240 miles) of the pole, the closest reached by man at that time, before giving up. The polar ice drift was now carrying them away from their goal, and it was time to turn homeward. Before leaving on their dash for the pole, the two men had prepared for their return. Their plan was to head southward over the ice to Franz Josef Land and from there to sail to Spitzbergen in two kayaks they had brought with them.

Their trip south was hindered by stretches of open water, forcing them to overwinter in a rude hut they built on an island (known today as Jackson Island) in the Franz Josef group. The following spring, after setting out in their lashed-together kayaks, they had the incredible good fortune of meeting a British explorer, Frederick Jackson, who took them back to Norway on his ship, reaching Vardø on August 13, 1896.

Meanwhile, Sverdrup and his crew of 10 drifted with *Fram* in pack ice that seemed bent on destroying the small schooner. Huge floes two metres (seven feet) thick squeezed the ship's sides in an icy vise exerting uncounted tonnes of crushing pressure. But the ship's designers had known their trade. *Fram*'s hull had purposely been built somewhat like a salad bowl, with smooth, sloping sides that defeated the powerful thrust of the floes by causing the ship to rise up on the ice rather than be crushed by it. For 28 backbreaking days, Sverdrup and his crew bored and blasted a path through the last 288 kilometres (180 miles) of solid ice, finally manhandling *Fram* through the gelid waters to the open sea near Spitzbergen. Sverdrup and

Fram reached Norway seven days after Nansen at the port of Sjervøy. On September 9, 1896, a tumultuous reunion and celebration erupted when *Fram* triumphantly sailed into the harbour of Kristiania from which she had departed 38 months earlier.

− 2 −

SVERDRUP'S SECOND ARCTIC EXPEDITION

IN JUNE 1898, TWO YEARS AFTER RETURNING from her first Arctic expedition, *Fram* sailed again, carrying Sverdrup on the second Arctic exploration trip which is the main subject of this book. There was no doubt in anyone's mind about the ability of either the man or the ship.* Both had more than adequately proven their mettle on the first *Fram* expedition. What no one could possibly suspect, however, was the connection that Sverdrup's second Arctic expedition would have one day with the question of Canadian sovereignty.

After clearing Kristiania harbour on June 24, *Fram* sailed for Kristiansand on the southern coast of Norway, which was reached two days later. At 3:00 p.m. on the same day, after sending off some last-minute telegrams to his King and to the *Storting* (the Norwegian Parliament), Sverdrup put out to sea, setting a westerly course toward the southernmost tip of Greenland.

The ship was well equipped and staffed for scientific observation. In addition to Captain Sverdrup, *Fram* carried 15 crew members with a wide range of experience and expertise. Victor Baumann, age 28, was the second in command. He had studied electricity and was in charge of magnetic observations. Olaf Raanes, *Fram*'s mate, at 33 had served all his life as a fisherman

---

* Scientific work and family responsibilities prevented Fridtjof Nansen from taking part in the second *Fram* expedition.

SOURCE: OTTO SVERDRUP'S *NEW LAND*

The crew of Sverdrup's second Arctic expedition, prior to departure from Oslo, 1898 (left to right):
Back row: Jacob Nödtvedt, Peder Leonard Hendriksen
Middle row: Karl Olsen, Rudolph Stolz, Ivar Fosheim, Herman Georg Simmons, Ove Braskerud, Johan Svendsen
Front row: Per Schei, Gunerius Ingvald Isachsen, Victor Baumann, Otto Sverdrup, Olaf Raanes, Edvard Bay, Sverre Hassel

and seaman. Gunerius Ingvald Isachsen at 30 years of age was the expedition's cartographer. The modern-day weather station of Isachsen on Ellef Ringnes Island bore his name before it was closed down a few years ago. Herman Georg Simmons, age 32, was the ship's botanist. He distinguished himself in his subject in later years by writing several scientific treatises. Edvard Bay, zoologist of the expedition, was a Dane with previous Arctic experience. He was 31 years old. Johan Svendsen, age 32, was the ship's doctor and thus a very important member of the crew. Per Schei, the expedition's geologist, was 23. Peder Leonard Hendriksen, at 39, had considerable Arctic experience and was the only member of the crew to have accompanied Sverdrup on his previous *Fram* expedition. Karl Olsen was the ship's chief engineer at 32. Jacob Nödtvedt, second engineer, at 41 was the oldest member of the ship's company after Captain Sverdrup, who was 43. Ivar Fosheim, the expedition's hunter, was 35. Adolph Henrik Lindström, 33, was the ship's cook and steward. Sverre Hassel was the youngest member of the crew at 22, and had general experience as a ship's mate. Rudolph Stolz and Ove Braskerud, both 26 years old, were the ship's stokers. Of the

16 men who sailed from the shores of Norway on June 24, 1898, on board *Fram*, two would never see their homes again.*

The schooner on which these 16 men sailed was well suited to her task. Conceived specifically for service in the Arctic ice pack, she had proven the skill of her designers and builder on her first polar expedition. For her second trip, Sverdrup had *Fram* modified to his taste by her original builder, Colin Archer, a Norwegian of Scottish descent. Forward of her mainmast, a new closed-in deck was extended to the bows, greatly increasing the living and working areas below decks. A new keel was attached in such a way that if heavy ice slabs caught on it, it would be ripped away rather then compromising the safety of the ship. The stern had two wells accessible from the deck, one for the rudder, the other for the propeller, through which these could both be hoisted up for repairs. The *Fram*'s hull was built of fine Italian oak sheathed with greenheart, a hard, rot-resistant wood from the forests of British Guiana (Guyana, as the country is now known). Her masts were of lofty Norwegian pine.

Fram was 39 metres (128 feet) long and displaced 725 tonnes. Her beam width was 11 metres (36 feet) and she drew 5 metres (16.5 feet) of water. When her sails were furled, 220 horses of coal-fired steam power drove her through the water at a steady speed of 6–7 knots in calm weather with a light cargo.

She was certainly not a big ship, but her quarters were relatively comfortable. Ten of the crew had their own cabins, five to a side along the midsection of the ship. Granted, they were only about two metres by two metres (six feet by six feet), but they were private. Six seamen and stokers shared two cabins in the stern, each with three bunks, but there were rarely three men in either of them at any one time. A large salon amidships served as a mess hall and leisure area—it was even equipped with a piano. A smaller salon aft provided more living area. The cook worked his culinary magic in a tiny galley equipped with a coal-burning stove. For a sailing vessel of her time, her quarters were snug, but not cramped.

After rounding the southern coast of Norway, Sverdrup set his ship on a course almost due west across 2,700 kilometres (1,680 miles) of storm-

---

* Interestingly, one man that Sverdrup tried to get as a member of his crew, but who could not accept because he wanted to complete his engineering degree in Berlin, was Herman Smith-Johannsen. This man later came to Canada, in 1919, settling in Montreal where he became the well-known cross-country skier "Jackrabbit" Johannsen. He lived an active life until he died in 1987 at the remarkable age of 111.

tossed ocean, navigating toward Cape Farewell on Greenland's southernmost tip. The crossing was not easy. The first days created a sense of false security as favourable winds and weather sped the tiny ship toward the world's largest island. Then, freshening weather accompanied by a quartering wind set the *Fram* to rolling. Many on board, especially the scientists, were "landlubbers" and had yet to taste the punishment the north Atlantic can mete out to those who dare sail her furrowed, wind-whipped surface in small ships. A regular procession of pale (and sometimes green-looking) men made its unsteady way to the doctor's quarters seeking relief from headache, shivering fits, pains in the stomach—calling their complaint anything but what the malady really was—seasickness. The most effective cure for the ailment was of course steady, dry land, but none was available, so the ill had to face the reality of the second-best cure—a slow transformation from quivering landsmen into steady seamen that only time and patient suffering before the mast can bring.

On July 1, *Fram* and her crew sailed along the last vestiges of the European continent—the Orkneys, a peppering of some 70 islands and islets that lie just north of the Scottish mainland—that they would not see again for more months than they knew. Two days later, the first gale and truly high seas of the crossing raked the decks of the tiny vessel, causing it to dip its nose shudderingly into the deep troughs between the towering, storm-tossed waves. Below decks, chests, cases, buckets and assorted loose objects were sent flying first to windward, then to leeward as the tiny ship heeled first this way, then that, under the irresistible force of the wind-whipped sea. It was a long crossing with contrary wind and weather, but from the first, relations among the crew members were excellent.

On the afternoon of Sunday, July 17, after a night of uncomfortable rolling, the crew descried the luminous ice blink,* far away on the horizon to the west, that could only signal a vast expanse of inland ice—the ice desert and eternal snows that blanket all of Greenland, except for the narrow coastal fringe where all its people live. The next morning, the crew could clearly see mountain peaks on the southern tip of Greenland.

Along the entire desolate 2,900-kilometre (1,800-mile) stretch of coast on the big island's east side, only one Inuit village of any importance existed in Sverdrup's time—Angmagssalik—and its population numbered only

---

* The ice blink is the reflection of the shining white surface of the ice cap onto the bottom of an overlying cloud layer.

about 300. A cold polar current flows south along the east coast making it far less hospitable than the west coast, where virtually the whole of Greenland's inhabitants live. The current brings with it solid evidence of its Siberian origins—drift ice and trees. Siberian rivers transport fallen tree trunks to the Arctic Ocean where the polar ice drift takes them over the region of the pole to finally release them along Greenland's east coast, where currents carry them around Cape Farewell and north up the inhabited west coast. Trees for lumber were welcome debris indeed to the Greenland Inuit of Sverdrup's time, as the huge island is essentially treeless.

The wide, south-flowing river of broken sea ice extends several kilometres south of Cape Farewell. Into this random jumble of millions of hummocks, floes and small icebergs steamed *Fram*, shouldering aside the smaller chunks and carefully avoiding the larger ones until she could advance only with great difficulty through the compacted icy mass. One large floe attracted the *Fram*'s attention because it cradled a large, water-filled depression from which the crew filled the ship's freshwater tanks while moored alongside. The sun shone warmly on the captain of the *Fram* as he conned his ship from the height of the crow's nest in order to better discover the open leads through the ice that offered the easiest passage. Sverdrup waxed poetic in his contemplation of the icy parade drifting past his gunwales. No shape on earth went unrepresented there, he mused, seeing a church and spire drift by, a sleeping princess in snow-white garb, an Akvavit bottle on a platter, a bullock's carcass with four legs in the air, tables, sofas, chairs, a horse mushroom, heads of polar bears and wolves—an endless procession.

Sverdrup was not the only Arctic sailor to have his imagination stimulated by icy sculptures. Some 35 years later, one of his countrymen-turned-Canadian, Henry Larsen, who sailed the *St. Roch* through the Northwest Passage in both directions, would write:

> *At times the ice formations take the oddest shapes: towering cathedrals with magnificent spires, battleships, bears or dancing girls. I never tired of watching the ice in the Arctic twilight.*

For the next five days Sverdrup had ample time to indulge his icy fantasies while forcing *Fram* across the polar current until she finally broke out into clear water south of Cape Farewell.

The expedition rounded the tip of Greenland and on July 25 steamed north past Godthaab, the capital of the island, today known as Nuuk. Sverdrup had good memories of Godthaab, which he had visited during an

1888–89 expedition after he and Fridtjof Nansen had skied across the barren ice cap from east to west, being the first ever to do so.

On July 28, *Fram* called at the harbour of Egedesminde where the Royal Greenland Trade Service had assembled a number of Inuit dogs at Sverdrup's request. The settlement's superintendent had bought 36 from local Inuit, but 3 had been killed by their companions and instantly devoured, and 2 others had run away. The superintendent, knowing full well the importance of dogs to an Arctic expedition, generously contributed 6 of his best to make a total of 37.

For Sverdrup, an essential supporting pillar for polar research was dogs—of the right kind. He knew his dogs, and only one kind would do—Inuit dogs. Norwegian (or Swedish) elk dogs, of which he had a few on board, were "utterly useless as sled dogs." Siberian huskies he had known on the first Fram expedition with Nansen, but they were "not up to the standard of the Eskimo dog."

Although all dogs are descended from wolves, the point at which the Inuit dog diverged from the wolf is uncertain, with estimates ranging from 25,000 to 50,000 years ago. Whatever may be the truth, one thing is certain—anyone approaching an Inuit village where dogs are still used will inevitably be greeted by a lugubrious serenade of echoing howls that raise the hackles on the back of the neck as surely as if they were wolves.

Inuit dogs can work their hearts out for their masters. Dog handlers who know what they are doing keep their dogs continually on the edge of hunger when traveling, feeding them only after the day's work. Under this condition they work best. Some have claimed that a high number of winter deaths among dogs is due to the fact that they cannot take continual winter darkness. Sverdrup refuted this, saying that high mortality was due to underfeeding, lack of shelter and lack of clean conditions. Contrary to the practice of many, Sverdrup fed his dogs every day. Their daily meals were walrus meat one day, dog biscuit the next, and fish the third day when available, otherwise it was walrus and biscuit on alternating days. The dogs had their kennels for protection from the cutting wintry blasts, and clean conditions to prevent matting of their fur with consequent loss of insulating value resulting in frostbite. Sverdrup also gave his dogs exercise every day rather than keeping them tied up when they were not working. Under this regimen, his dogs fared well all winter.

Peter Freuchen, the famous Danish Arctic traveler and raconteur, mentioned an inconvenience when winter traveling with Inuit dogs. Defecating

SOURCE: OTTO SVERDRUP'S *NEW LAND*
Four of the expedition's dogs. From right to left: Gammelgulen, Lasse, Svartflekken and Basen

was difficult if there were any loose dogs around since, being on the sharp edge of hunger, they were extremely interested in the end product and could hardly wait for it to drop. Freuchen's method was to use the temporary traveling igloo, in which he had passed the night, for tending to this business before departing in the morning. He would then unblock the entrance and allow to dogs to rush in and, as he put it, "gulp down the delicious, steaming mouthful."

The weak and the wounded beware! Inuit dogs are quick to pounce and devour the underdog, literally. If two dogs get into a fight and one is downed and badly bitten, this may well seal its fate, for the rest of the pack will jump in, snarling and snapping, and in a matter of minutes all that may be left of the unfortunate one is a few tufts of hair. When a pack senses weakness, it is virtually impossible to stop it until the victim is devoured or seriously maimed, be it canine or human.

Father Charles Choque, in his recent book Mikilar, recounts the following tragedy which occurred in 1925 in Chesterfield Inlet:

> *The wife of Sergeant Clay of the Canadian Mounted Police fell prey to a pack of dogs who chewed her leg leaving the bone*

> *exposed. Her husband was [away]. An Eskimo lady ... ran to alert the two policemen who had stayed home and they rushed to the woman's side. The ferocious beasts were by then uncontrollably excited at the sight of blood. With sticks the policemen succeeded in warding off the dogs. They carried the poor woman home and gave her first-aid but nobody knew exactly what to do. The only solution seemed to be the amputation of the leg. The next day, at the request of the patient, who was writhing with pain, Father Duplain accepted to perform the operation. Thanks to a good dose of chloroform and much composure all went well. She was out of danger—so everyone thought. But one ill-advised good Samaritan gave her water to drink. She became delirious and died. What sad news for her husband.*

In Sverdrup's words, the Inuit dog "has the persistence and tenacity of the wild animal, and at the same time the domestic dog's admirable devotion to its master. It is, so to speak, the wildest breath of Nature, and the warmest breath of civilization."

On July 29, *Fram* departed Egedesminde bound for Godhavn on Disco Island, reaching it at 3 o'clock in the morning the next day. Thirty-five more dogs were taken on board as well as 36 tonnes of coal and fresh water. From August 2 to August 4, the expedition sailed north to Upernavik to pick up another 30 dogs. However, there had been an epidemic of infectious disease that winter among the Upernavik dogs, and upon learning this Sverdrup decided that no chances could be taken of spreading it to the healthy dogs already on board.

Vesta, one of the bitches, gave birth to five puppies at this point. A few days later, Sussaberet also produced five puppies. The idyllic scene was shattered one day when Vesta suddenly snapped up one of Sussaberet's pups and instantly devoured it, confirming the Inuit dog's heritage from the wild. The two families were kept separated from that point on.

For two days after leaving Upernavik, the *Fram* steamed northwards in brilliant sunshine at a steady five knots over a glassy smooth sea, until the expedition was inside Melville Bay where many vessels have come to grief in the notorious crushing ice for which this body of water is known. On August 7, *Fram* entered compact ice and towards evening, thick fog came rolling in effectively immobilizing the ship for the next several days. Sverdrup finally decided to force his way through by starting up *Fram* "at full speed, boring and shouldering herself a passage, brushing one ice-floe here and another there." On August 16, *Fram* found open water and the expedition headed north for Etah, killing three walrus on the way for dog food.

At Etah, high up on the northwest coast of Greenland, there had once been a sizable community of polar Inuit, but by Sverdrup's time, the village had been abandoned in favour of Thule, where the Danes had established a trading post. The site of the old village was still used by polar explorers as one of several places to leave records and messages about the progress of their expeditions. In this way, if an expedition was overdue, rescuers could at least narrow down the search based on the last place a message had been left and what it said about the expedition's next destination. As well, if an expedition were lost, at least a partial record of its work to date would be preserved.

After leaving its own records and messages in a cairn, Sverdrup's expedition continued north along the coast of Greenland, but on the evening of the August 17, they had to beat a retreat back toward Etah due to packed ice and an unfavourable wind. On their way south, they passed Lifeboat Cove, where the crew and passengers of Captain Charles Hall's *Polaris* had run into trouble on October 16, 1872.

Hall's ship was stranded in the ice in a sinking condition with 14 men aboard. The day before, the rest of the crew and passengers, 12 men, 2 women and 5 children including an Inuit mother with a two-month-old baby had been moved onto the sea ice along with most of the provisions and clothes in case the ship foundered. Towards evening, a strong wind arose. At around 10:00 p.m., the force of the wintry blast broke the *Polaris*'s moorings, carrying away the ship and her 14 crew members.

The fate of those on the sea ice also took a turn for the worse, when a huge ice floe—on which they were resting—broke away from the main ice pack and drifted away. All winter long, the 19 people marooned in this ice floe prison drifted south, kept alive by two Inuit hunters among them who, having kayaks, harpooned seals whenever they could. On April 30, more than six months after drifting away from the *Polaris*, they were finally picked up off the coast of Newfoundland, some 2,400 kilometres (1,500 miles) from their starting point. All had survived. In fact, far from diminishing, the population of the ice floe actually increased during its southern drift when an Inuit woman gave birth.

The 14 men caught on the drifting and sinking *Polaris* when the ice floe broke away had been blown into Lifeboat Cove, where they spent the winter. In the spring, they built two boats out of the wood of their ship and rowed south along the coast of Greenland to Melville Bay where they were picked up by a whaler on June 23, eight months after *Polaris* had drifted off into the night. All 14 who had been stranded on the ship also survived.

When the *Fram* was abreast of Etah once more, Sverdrup crossed over to Ellesmere Island and steamed north along its coast, still trying to force *Fram* into Kane Basin and accomplish his first objective—to go as far north as possible, then exploring the extreme north coast of Greenland by dog-team, and as far down the east coast as he could. It was only August 17, but already impenetrable masses of ice and contrary winds combined to force Sverdrup to anchor his ship in the ice pack north of Pim Island. The captain then began a game of "chicken" with the elements. He still wanted to force Fram to the north, but he had to be very careful not to let his ship be locked into the ice to spend the winter in the open, exposed to every buffeting wind and movement of the ice. A strong north wind forced him to take refuge in Rice Strait, between Pim Island and the mainland of Ellesmere Island. Here, on August 21, Sverdrup and his crew finally ceded victory to the elements and sailed Fram into a sheltered little bay which became the expedition's first winter quarters. He named it Frams Havn (Fram Haven).

- 3 -

THE FIRST WINTER

WINTER STARTS EARLY IN THE HIGH ARCTIC. It can be a dangerous time, with temperatures that plunge to awesome depths during the long, lonely hours of polar darkness. Powerful blizzards shriek across the land for days at a time, causing all animal life to seek shelter from the cutting blast, essentially putting a temporary end to normal activities of life such as traveling and eating. It is an unforgiving land that does not easily suffer fools.

It was not even the end of August and *Fram* was already in her winter quarters, unable to proceed further up the coast of Greenland according to plan because of pack ice. The temperature was not yet very cold for that latitude, ranging from about -10° to +5°C (15° to 40°F). Preparations had to start immediately, however, in order to be ready for the bitter, lonely, black cold of midwinter, when temperatures could often drop to the -40s and -50s (similar in both Celsius and Fehrenheit). Two things were absolutely essential, without which death would be certain—adequate food and proper shelter. Securing a food supply came first. Making *Fram* snug against the deep penetrating cold of continual night could wait—for now.

Sverdrup had provisioned his ship for at least four years. But there was one absolutely essential commodity he could not bring from Norway, not three or four years' worth, anyway, and that was meat. The sea and the land would have to be harvested for meat, both for dogs and for men.

As soon as Sverdrup realized that *Fram* was blocked from sailing further north that year, he set his men to walrus hunting. Many tonnes of meat were needed for the ravenous dog teams.

Walrus hunting is not a sport for the faint of heart. An adult male can weigh 1,250 kilograms (2,800 pounds) and measure up to 3.7 metres (12 feet). Females are three-quarters the size of the males—still a formidable and extremely dangerous prey. Both have oversized upper canines in the form of long tusks that are useful not only for digging up clams on the seabed, but also for slashing to ribbons any hunter unfortunate enough to fall in the sea among the huge animals. Walrus were hunted from open boats, painted red inside to blend in with the blood of the stricken animals and white outside to make them as hard as possible for the animals to detect among the ice floes. The boats were robust and equipped for an absence from the ship of two or three days up to a fortnight. One never knew when fog, gale or ice, or a combination of these, would impose a long absence from the security of the mother ship. Boats carried a cooking pot, a coffee kettle, a small chest of provisions, an oaken fresh-water barrel, a hatchet for chopping wood, a heavy, strong ax for removing tusks from the walrus, lances for dispatching the animals, block and tackle to haul them onto the ice for skinning, seal hooks, two wooden harpoon shafts and eight harpoons with 3–4 metres (10–12 feet) of line. The boats were crewed by four men—one harpooner, one coxswain and two rowers.

When a herd was sighted, the men bent their backs to the oars with all their strength. They dug them in, pulling against the icy waters in a strenuous outpouring of effort to catch up with the animals and give the harpooner his chance to hurl his steel-tipped weapon deep into the back of the nearest beast with the fullest force of both his arms, shouting the Norwegian equivalent of "Got fast!" in a cry of triumph. The rowers immediately shipped their oars, for now the boat had a powerful "motor" towing it along with the herd as the wounded animal struggled not to be left behind by its mates. Spray flew from the bow as the boat raced along with the herd amid the tremendous uproar of excited barking and trumpeting of the now fully alarmed beasts. The coxswain steered the craft as best he could so that one of the rowers could hook up the passing wooden harpoon shaft that had come loose from the steel tip when the animal was struck. If it was missed, the poor coxswain received all the blame—there were normally only two shafts in the boat, and they had to be reused repeatedly, for the harpooner would soon be thrusting out his remaining seven steel weapons into the backs of seven more unfortunates. Each time, the loose wooden shaft was hooked up from the sea as it flashed past the speeding boat. The first animal was hauled astern, dispatched with a lance, and had a loop cut into its neck-skin through which a rope was passed to tie it to the boat. The harpooner cut out the steel harpoon head, jammed it onto a recovered wooden shaft, ran to the bow and again thrust his cruel weapon home. "Got fast!" Then back to the stern where the second animal had been dispatched and the harpoon cut out, to be jammed onto a wooden shaft and thrust into the next victim. With only two wooden shafts and eight steel harpoons, almost a whole herd could be secured. The limit came when the wounded walrus could no longer tow the boat and its trailing carcasses along fast enough to keep up with the remaining animals.

Walrus have been known to exact their revenge on their tormentors by swimming under a boat, turning on their backs, and punching a hole in the bottom with their extremely hard heads. Infuriated males can hook their tusks over the gunwale of a boat and spill the flailing crew into the bloody maelstrom of maddened and slashing animals.

Even Sverdrup remarked that this was a hell of a way to treat animals, not to mention men. The water surrounding the boat was blood red—the men were smeared with blood, the inside of the boat was slippery with it. Killing the animals was but half the work. Next came the drudgery. The hunters rowed toward the ice, towing their meaty prize, for the arduous task of hauling out the carcasses to gut and clean them. After hours of stinking, bloody work, the hunters boated the empty carcasses over to Pim Island, about 1.5 kilometres (one mile) from Fram Haven, and lay them on the *kjöthaugen*, or "meat heap," as Sverdrup christened the growing pile.

Meat for the men came from the land in the form of muskoxen. These members of the bovine family can weigh up to 400 kilograms (900 pounds) and, as Sverdrup's men found out, the beef they provided was of superior quality and taste. Inland, a day-and-a-half by sled from the ice-imprisoned *Fram*, Sverdrup established a semi-permanent camp to serve as a jumping-off spot for hunting trips and journeys of exploration over the bleak and deeply frozen surface of Ellesmere Island. The camp was christened Fort Juliana when the cook, asked by Sverdrup to make some Julienne soup, did so, calling it "Juliana soup."

It was here at Fort Juliana, on October 6, that Sverdrup and Bay had a disturbing encounter. The two men were brewing a pot of coffee when they spied a black spot in the distance slowly moving up the fiord toward them along their sled track. With time, the spot resolved itself into an Inuit sled, drawn by eight straining dogs and bearing two fur-clad men. Sverdrup had a good idea who it might be and looked forward with anticipation to see if he was right. He was.

One man was an Inuit driver and the other was Robert E. Peary, the famous American Arctic explorer, whom Sverdrup knew to be in the area. Peary also knew that the Norwegian was in the area, and he asked him if he was indeed Sverdrup. The strange and disturbing part of the visit was its shortness and curtness.

It was common courtesy in the Arctic when a camp was visited by a sledding party to offer the often cold and bone-weary travelers cups of steaming-hot coffee and food to revive them. Neglecting to do so was gauche in the extreme according to the code of the North. Refusing to accept also showed a great lack of courtesy. After all, these encounters did not happen every day. Refuse, though, is exactly what Peary did, claiming that his tent was a mere two hours drive away and that he was on his way home to dinner. This was strange behaviour indeed. Peary's refusal, coupled with the fact that he never mentioned this encounter in his exploration records, nor two subsequent meetings between the two parties, provides an insight into the mindset of this famous man.

Robert Edwin Peary was nothing if not singled-minded, with Arctic exploration his one, all-consuming interest. Exactly when he had set his sights on the discovery of the North Pole as his life's ambition is not known, but when he met Sverdrup in 1898, he had already made his first unsuccessful attempt during the winter of 1893–94. Peary was aware that Sverdrup's first *Fram* expedition with Nansen in 1893–96 had also been an attempt to reach the coveted pole. Sverdrup's present mission had nothing to do with reaching the pole, but this, Peary could not know. He obviously considered the presence of Sverdrup in the same area as a threat to his long dreamt-of goal, which he is generally credited with reaching in 1909 (although his claim is still challenged in certain quarters, even today). Sverdrup was an unwelcome intruder in Peary's garden, and Peary's resentment was unbridled.

To the Norwegians, however, "Peary's visit was the event of the day.... We talked of nothing else, and rejoiced at having shaken hands with the bold explorer, even though his visit had been so short that we hardly had time to pull off our mittens."

Sverdrup learned that Peary's ship, the *Windward*,* had arrived in the

---

* The captain of Peary's ship was John Bartlett of the well-known seafaring family from Brigus, Newfoundland. The ship's first mate was his nephew, Robert Bartlett, soon to be famous as Captain Bob Bartlett, who later captained Peary's *Roosevelt* in 1909, when Peary made his much-disputed discovery of the North Pole.

area a few days before *Fram*, and had been beset in the ice some 100 kilometres (60 miles) north of Fram Haven as early as August 18. Sverdrup could not know it at the time, but before sailing for the Arctic that year, Peary had learned of Sverdrup's plans to be in the same area at about the same time. He had sped up his plans and forced *Windward* as far north as possible into the ice pack before winter set in, to establish his precedence in the area.

During September and October the Norwegian party busied itself establishing a muskox meat depot west of Fort Juliana. Sverdrup found that neither the meat nor the milk of the animals displayed any hint of a musky flavour or odour, so he rebaptised them "polar-oxen." If hunting walrus was always fraught with danger and high-level excitement, this was not usually so with polar-oxen. Sverdrup wrote that hunting these animals, which he aptly describes as reminding him of funereal horses with trappings reaching the ground,

> *is not sport: it is plain butchery; it requires little skill, and causes one no excitement. Anybody can set a team of dogs on the trail and then quietly follow them with his gun, walk up to the animals, and shoot down the whole herd. But what could we do? Meat we must have, and necessity knows no law.*

Muskoxen have a characteristic way of defending themselves that may serve them well against wolves,* but which is completely useless against man:

> *... the dogs had winded the oxen. I undid the connecting lanyard and let them head up the slope, followed by Bay, who scrambled after them—I liked looking on at this sort of sport.* [This sentiment of Sverdrup's contrasts sharply with the sympathetic one expressed above *vis-à-vis* these animals.] *When I reached the first slope, I observed that the 4 oxen were standing in wait for the dogs which were making toward them. It was evidently their intention to give battle, and when the dogs came up a curious scene ensued.*

---

* Wolves are the only natural predator that muskoxen must fear. It is not the adults that are vulnerable to attack by wolves, but the young, and the oxen's defense is admirably suited to preserving the young from wolves. But when the predator is modern man with his efficient killing machines, standing around in a circle is a recipe for extinction. The species was indeed very nearly wiped out before protective laws were put in place and enforced in more recent times.

> The [four] oxen had formed a square. They stood at regular intervals from one another, with their hindquarters together and their heads outwards. Then, in turn, and with lightning speed, each one made an advance in the shape of a circular movement from left to right. At the same moment that an ox regained his place, his neighbor on the right sped out on a similar attack, and thus they went on uninterruptedly, with almost military precision. As long as the maneuvers continued, one of the oxen was always out on a movement of attack, endeavouring to rip up one or more of his adversaries.
>
> The size of the attacking circle always seemed to be determined by the distance of the enemy and the nature of the ground. As a rule, the animals advance ten or twelve yards [9–10 metres] from the square, but once I saw them make attacks to a distance of a hundred yards [90 metres]. The remaining oxen always cover the gap in the square, but immediately make room for their comrade when he returns from his round. Now and then, when the battle is a long one, they stop, and then begin again with renewed vigour. The greatest degree of precision is attained by oxen of the same age. Like old combatants, they seem to thoroughly enjoy defending themselves, and appreciate the sporting element in it.
>
> I have seen herds of as many as thirty animals form a square, with the calves and the heifers in the middle, and the bulls and cows standing in line in defense at distances as equal as the points on the face of a compass.

While hunting muskoxen was not generally a sporting proposition, on the other hand, one could not be too complacent. Sverdrup's hunters occasionally met with animals that did not behave as expected, with resulting moments of sheer panic and grave danger quite equal to those experienced when hunting walrus:

> A large ox was still standing a little distance off, so I let go the dogs onto it, and left Simmons and Schei to do the rest. It was their first experience of polar-ox shooting. They followed the animal a little way along the flat-topped ridge of sand and then Schei dropped behind a stone, from which he meant to get a resting shot. Simmons was just standing, wondering whether he

> *should do likewise, but before he could make up his mind, the ox set off at full gallop down the slope, stones and earth flying from under its hoofs. It headed straight for the discomfited sportsmen, and so extraordinarily quick was the animal, that not one of the dogs could keep up with it.*
>
> *It could not have been pleasant to be Simmons or Schei at that moment. It was difficult for either of them to shoot, for if they missed they might hit a dog; in any case, to shoot resting was impossible. This Schei also perceived, and he started up to aim; but the ox advanced on him so rapidly that he was not ready for it in time. The same was the case with Simmons—he had got a cartridge jammed—and now there was only one thing left for them to do: run to one side and avoid being tossed.*
>
> *I had my own reflections on the subject as I stood looking on at the performance, but they were of short duration. At the mad pace at which the ox was going it was impossible for it to remain up under the boulders, and so down it came, heading straight for me. Behind the ox were both the shooters and the dogs, and if I missed, one or other of them might be killed. There was no time for hesitation, so I sent the ox a bullet at twenty yard's [18 metres] distance, but without its having the slightest effect. It rushed straight at me with the same furious pace as before, and there was absolutely no possibility of getting in a new cartridge. I had to do what my comrades had done. The animal flashed past, but, my second shot being ready just as it was turning round, I gave it a charge which hit it on the back of the head. It fell on the spot.*

Sverdrup does not reveal his nor his companions' emotions at this sport, but it's safe to assume that blood pressures rose and heart beats accelerated.

Now that the meat for both men and dogs had been secured, shelter against the deep black cold of the Arctic winter was the next order of business. *Fram* was basically a sound home for the expedition to spend the four sunless months ahead, but her deck construction was not efficient in keeping out the Arctic cold. On November 1, preparations for overwintering began. First, tarpaulins were stretched over the cold-leaking skylights and a wall of snow a metre (a yard) or so thick was raised around them as a windbreak. Then the part of the decks above the cabins was paved with insulating snow. Even the smallest crack had to be sealed because of a peculiar characteristic of Arctic snow. The Arctic is virtually a desert. Very little snow falls there. What snow there is blows back and forth across the bleak, rocky surface, driven by howling blizzards and unobstructed by trees. The points of the star-shaped snowflakes are broken off by this rough treatment and the flakes turn into tiny, truncated little balls of frost that slip into the finest cracks. Blizzards have been known to blow sizeable snow banks into houses through the hairline crack between a door and its jamb.

Next, kennels were constructed for the dogs out of blocks of ice. Each team had its own kennel, with wooden partitions to separate the dogs and prevent fighting. The importance of these precautions is illustrated by the fate of Fin, one of the Norwegian elk-dogs which had the misfortune of being attacked by a team of Inuit dogs one day. By the time the men pulled the snarling mob apart, the only thing left of poor Fin was the tip of his tail. The rest of him had been torn apart and devoured.

Sverdrup was satisfied with the progress made by his men:

> When all this had been done, we considered everything very warm and comfortable and that we might anticipate the darkness of winter with equanimity, at any rate we thought we should not be cold.

Although the crew of *Fram* would not be physically cold, however, the psychological chill of the deep, dark Arctic night would certainly seep into their hearts. On October 16 the sun dipped below the horizon, not to appear again before four months had passed:

> We were looking at the sun for the last time that year. Its pale light lay dying over the inland ice; its disc, light red, was veiled on the horizon; it was like a day in the land of the dead. All light was so hopelessly cold, all life so far away. We stood and watched it until it sank; then everything became so still that it made us shudder, as if the Almighty had deserted us and shut the gates of Heaven. The light died away across the mountains, and slowly vanished, while over us crept the great shades of the polar night, that kills all life.
>
> I think that each of us, standing there, felt his heart swell within him. Never before had we experienced homesickness like this, and

little was said when we continued on our way. For a few days longer we were able to see a faint light on the highest mountains at noon—a suspicion of dawn in the south which told us that there was still life to be found somewhere in the world. Then even that was gone; we had entered on the great night.

What might not these coming four months of darkness bring? Things so terrible had occurred up here in the polar night that they might well make anyone pause and think. Here came Franklin, with 129 men. The polar night stopped them; not one returned. Here came Greely, with twenty-five men; six returned. A year or two after Nordenskiöld wintered safely at White Bay on the Siberian coast, many men of De Long's expedition died of scurvy on the same coast, in the midst of a superfluity of food. And yet in spite of all that had happened, in spite of all the horrors that had been experienced, we felt, on the whole, secure.

Until mid-November, expedition members continued making small sled journeys into the deepening cold, shorter and shorter each time until finally they quit altogether and their period of hibernation began. But it was not a period of idle hibernation. The ship became a centre of earnest preparation—preparation for Sverdrup's planned exploration of the northern coast of Greenland the following year.

Nödtvedt built a forge a short distance from the ship, protected by walls built of blocks of snow. Here, in the welcome warmth and glow of his forge in the deep Arctic night, the blacksmith turned out all manner of iron and steel implements—knives, hatchets, bores, crowbars. On board ship, Baumann plied his tentmaker's skills stitching canvas shelters for the future explorers. Raanes, the mate, sewed kayaks. Braskerud was a general handyman—both a carpenter and a watchmaker. Isachsen and Bay toiled in the galley pressing dog food into moulds to ease the feeding chores. Fosheim built the hut to house the four men who would winter on the northern tip of Greenland the following winter, as well as the sleds that would carry the travelers and their gear into unexplored lands. Svendsen, the doctor, and Hendriksen, his scientific assistant, went off every two weeks to take water temperatures, faced with the formidable task of boring a hole through two feet (half a metre) of ice each time until a seal appropriated the opening for its breathing hole and kept it open for them. The seal was rewarded with a fish each time they took readings to encourage it to keep up its much-appreciated duties.

The dark winter period became a time of febrile preparatory activity, punctuated with periods of respite to celebrate birthdays and holidays. Christmas was a very special period during which work ceased until after the new year had begun. The ship was polished and made to shine: flags, pennants and Japanese lanterns were hung to counteract the frozen darkness outside and create an atmosphere of cheer. On Christmas day, Sverdrup and his crew treated themselves to a banquet supper, complete with coffee and liqueurs and special Norwegian cakes for dessert. Later, champagne and hot grog warmed up the evening. It was with deep feelings of nostalgia that each toasted his dear ones back home, but it was best not to dwell too long on their homeland where they knew light and life still existed, lest a melancholy homesickness set in. Presents were exchanged, and dancing and singing carried on long into the frozen Arctic night, as the crew created its own small oasis of warmth and cheer in the cold and cheerless black emptiness surrounding them.

January and the early part of February were dull and routine, with nothing much to excite the spirit except the perceptible, gradual lightening of the sky at midday. On February 12, Olsen, the ship's engineer, and Svendsen, the doctor, set out on a day trip to try and find signs of Greely's last camp.

In 1881, Lieutenant Adolphus Greely and 24 men set out to establish an American post for observing magnetic and meteorological phenomena at Fort Conger on northeastern Ellesmere Island. It was the first International Polar Year, sponsored by 11 countries, of which a number established northern observation posts. Thus started one of the most gruesome and pathetic chapters in the annals of polar exploration.

The plan was that Greely's post would be relieved in 1882 and 1883 by ships bringing supplies and fresh replacements for the men. The ice conditions did not allow relief ships to reach the American post in either year. In fact, the 1883 relief ship *Proteus* was nipped by the ice and sank. By August 1883, Greely and his men knew that they could not stay another winter where they were, since their food reserves were virtually exhausted. They left Fort Conger in four small boats and worked their way south through Kane Basin, picking up small caches of food along the way, but all along the coast of Ellesmere, they searched in vain for the relief ships and the much-needed provisions that they might have left. They turned their boats north again—they were down to two boats, the others having been destroyed by the ice—because of an arrangement that had been made for news to be left on a small island* about the location of the deposits left by the ships. But in fact, no ship had been able to leave any, as Greely was shocked to learn when nothing was found on the island. Greely and his men moved onto Pim Island and set up a desperation camp. They built a house of stones, using their last boat for a roof and eked out a hopeless winter of starvation. They were able to capture only a few seals and a small bear. Leather from their clothes and footgear was cut into strips, boiled up and eaten.

During the terrible winter, 14 died of starvation, one of scurvy, one frostbite and exhaustion on a sled journey, one drowned trying to obtain food and, finally, one was shot on Greely's orders because he stole food belonging to others. One poor man who did not die lost all his fingers, both his feet and his nose to the deep Arctic frost. A spoon was tied onto the stump of his arm so that he could feed himself.

In 1884, a fleet of three relief ships steamed up the strait between Greenland and Ellesmere looking for the expedition. On the northern part of Pim Island they saw a human form standing on a small hill. The figure stumbled and fell down the hill toward them, a pathetic, sunken-cheeked ghost with wild eyes and horribly matted beard and hair. Up past the hill stood Greely's tent where a gruesome sight awaited the rescuers. The desperate survivors had had to abandon the stone house when the spring thaw let a deluge in through the roof. Inside, near the door lay an emaciated form with staring open eyes—dead. On the opposite side of the tent lay another, with no hands and no feet, a spoon tied to the stump of his arm—just barely alive. In his tent, on hands and knees, a horribly weakened Greely with a long, dirty beard and wide open eyes tried to look up at his saviours and whispered, "Here we are, dying, like men." Some 45 metres (50 yards) from the tent, 10 men lay in their graves. One lay unburied at the foot of a hill. Four had been laid near shore and washed away by the waves. The body of the food-thief who had been shot sprawled out on a patch of snow near the tent. When the bodies were being prepared for the trip home, it was found that the flesh of six of them had been partially removed.

---

* The island was Brevoort Island, just off larger Pim Island which was only a few kilometres from where *Fram* was iced-in for the winter 15 years later.

When Olsen and Svendsen tried to find the spot of Greely's terrible ordeal, Olsen came close to sacrificing his own life to unlucky Pim Island. He was out of shape, having done virtually no exercise the whole winter. On the return trip, he tired quickly. While still a few kilometres from Fram he could go no further. Svendsen left him and ran for the ship, knowing that he had to hurry if Olsen were not to freeze to death. Sverdrup and several of his men hurried to their stricken friend with a sled, bundled him up and sped him back to the reviving warmth of the ship. They stuffed him into his bunk, filled him with hot chocolate and gave him a bit of food to eat. Olsen recovered surprisingly quickly, and the following day he was almost as good as new.

In mid-February, the first glow of the returning sun painted the tops of the mountains with a wash of yellow:

> *It was like drinking in life with our eyes, like growing ten years younger, when the first clear sunbeams gilded the slopes and the sun stood above the crest of the mountains for the first time. Every man came on deck beaming with delight.*

Although the first sun of February warmed the heart and soul, it is a fact that the coldest temperatures of winter in the Arctic are registered after the sun's reappearance. On February 22, the thermometer plunged to -55°C (-67°F) at Peary's ice-imprisoned *Windward*, just 100 or so kilometres (about 60 miles) north of *Fram*. On March 7, three members of Sverdrup's crew set out on a sled trip along the coast of Ellesmere to the north. When they returned four days later, Schei's feet were frozen. Three toes on one foot and two on the other had turned black and had to be amputated. One still had to treat the Arctic with respect.

In the middle of March, a strange team of dogs with one fur-clad man aboard approached *Fram*. A Greenlander with eight dogs was driving north to Peary's *Windward* to tell some of his Inuit of the drowning of a number of their relatives while walrus hunting. Baumann and Hassel took advantage of the opportunity to visit the great American explorer at his icebound ship and accompanied the lone Inuit on the remaining two days of his trip north. Peary had frozen his feet on a sled trip at the end of February and nine of his damaged toes had been lost to the knife. He was confined to his berth at the time of Baumann and Hassel's visit. Baumann's account of his meeting with Peary leaves the impression that their reception was cordial, but certainly not to be noted for its warmth. One can only guess how much

SOURCE: OTTO SVERDRUP'S *NEW LAND*
The Fram *in Fram Haven, all decked out for the Norwegian national holiday, May 17, 1899.*

of Peary's aloofness was due to his state of health at the time and how much was due to the "intrusion" of Sverdrup's expedition into what he must have considered his personal territory. When Baumann took his leave, Peary's parting words were that he could not make any promise to visit *Fram*, either for himself or his companions. Sverdrup's presence in the area apparently still weighed heavily on his mind.

By April, the cutting edge of the deepest winter frosts had been dulled by the reappearance of the sun. It was a time for exploration of the land by dogsled. Sverdrup and Bay left *Fram* on April 17 for an east-west crossing of Ellesmere, driving to the end of Flagler Fiord, then overland and down to a new fiord on the western coast of the big island never before seen by modern Europeans. Sverdrup named this sinuous finger of sea-water Bay Fiord in honour of his traveling companion. Looming in the misty distance, at the seaward end of Bay Fiord, the two men saw a large land mass to the west. In the following years, the expedition would explore this land, discovering and naming it Axel Heiberg Island in honour of the Norwegian consul who was one of Sverdrup's sponsors. But for the moment, the shadowy mass at the far end of the fiord remained a beckoning mystery as Sverdrup and Bay retraced their steps to their ship, arriving there on May 6.*

---

* This was not the first time Ellesmere had been crossed, but it was the first time by the route now known as Sverdrup Pass. In 1883, Lieutenant J.B. Lockwood of the ill-fated Greely expedition had crossed Ellesmere 240 kilometres (150 miles) further to the north.

SOURCE: OTTO SVERDRUP'S *NEW LAND*
Inuit visitors, spring 1899, Fram Haven.

During May and June, Isachsen and Braskerud also crossed Ellesmere by dogsled, this time over glacier ice further to the south. These two crossings of Ellesmere were the major explorations of the 1898–99 winter period—an achievement that would pale in comparison to the trips of discovery of the expedition's subsequent years.

While Isachsen and Braskerud were away on their crossing, an event occurred among the members remaining with the ship that would have considerable significance for the expedition, but especially for Braskerud in the days to come.

On June 2, two expeditions of two men each set out together from *Fram*, splitting up into two groups at Outer Island in Hayes Sound. Sverdrup and Simmons entered Flagler Fiord to explore the valley at the head, Schei and Svendsen, the expedition's doctor, headed south into Beitstad Fiord.

Ice conditions in Flagler Fiord were bad, preventing Sverdrup and Simmons from going very far. They retraced their steps, stopping to camp at Fort Juliana on June 6. Soon after their arrival, they were surprised to hear voices outside their tent. It was Schei and the doctor who had also turned around because Svendsen's eyes had become irritated by the snow glare and also, more ominously, he had suffered from chest pains.

© JERRY KOBALENKO
Johan Svendsen's cross, Fram Haven.

Svendsen rested for a day and felt much better. Sverdrup offered to take him back to the ship, but the doctor refused, saying that he would be all right by himself while Sverdrup, Simmons and Schei returned to explore Beitstad Fiord for a few days. When the threesome came back a few days later, a grisly sight awaited them—Svendsen was dead, shot by his own hand, a suicide note by his side.

The doctor had had a problem. Sverdrup had known about it, as had other members of the expedition, but they had only learned about the problem after leaving Norway. Had Sverdrup known of it before, he would not have accepted Svendsen as a member of his expedition. Svendsen was addicted to morphine, a hazard of his profession as doctor. He had tried to kick his habit, asking Sverdrup to hide the keys to his medicine chest. This, the captain refused to do, saying that the doctor was a grown man and had to deal with this problem himself. Svendsen probably tried hard to overcome his addiction—Sverdrup mentions that the doctor had not brought any morphine with him on this last exploration trip of Beitstad Fiord with Schei. But Svendsen's heavy load was too much for him to bear, and his note indicated that he could no longer stand the loneliness and isolation of this Arctic journey from which he could not escape. He solved the problem in his own desperate and lonely way. This was a serious blow for the expedition, not yet through the first year of a trip planned to last three. Now, the doctor was dead. It was also a deep personal loss for the rest of the crew.

It was a sad cortege that sledded back to *Fram* with Svendsen's body to announce the news to the rest of the crew. As Sverdrup noted, "Svendsen was a great favorite with us all, and I know this will be a heavy blow to the fellows on the *Fram*." In spite of his problem, the doctor had been a lively and appreciated member of the small society.

A gloomy pall descended on the little ship so far from home, so lonely, so sad. On June 16, Svendsen was given a seaman's burial, slipping forever into the frigid Arctic waters of Rice Strait through a hole chopped in the ice. Hymns were sung by the grieving crew, and the Lord's Prayer was recited. A wooden memorial cross was erected on a hill overlooking Svendsen's last resting place.*

Sverdrup's mood remained somber. On June 24, in spite of the awakening world of spring surrounding him, in spite of the midnight sun, he wrote in his diary,

> *It seems as if this expedition is doomed. First we were stopped here last autumn. Then the doctor died. Isachsen and Braskerud have not yet returned from the inland ice and no one knows how they are faring; we must hope that they at least, are in no danger, but it would be bad enough for the expedition if they found it necessary to shoot some of the dogs. Added to this the condition of the ice in Kane Basin looks so unfavourable that a miracle will have to happen if we are to get up there this year.*

Sverdrup's ominous premonitions were not without foundation. Isachsen and Braskerud returned on July 2 after successfully crossing Ellesmere on the ice cap, so that part at least turned out well. The ice in Kane Basin, though, was another matter.

On July 24, *Fram* finally slipped her icy fetters and moved out from Fram Haven, her home for the past 11 months, to tackle the thick, closely packed polar ice of Kane Basin. The grinding, heaving ice was impenetrable and extremely dangerous. The ship had to beat a hasty retreat back to Fram

---

* Svendsen's cross was visited in 2001 by Jerry Kobalenko, author of *The Horizontal Everest*, and found to be still keeping its lonely vigil, although the lettering on it was now very faint.

Haven before getting into serious trouble. On August 4, another attempt was made with partial success. Then the crew sighted a dark smudge on the horizon ahead that sent an ecstatic thrill through their hearts. Smoke! Another ship! News from home! The masts of another steamer rose above the horizon as the two ships approached each other. The newcomer was a supply ship from America that was coming to replenish the stores of Peary's expedition on board *Windward*, some distance to the north. Their semaphore signals indicated that they had mail for *Fram* and her crew.

However, the letters would have to wait, for "some four or five miles [6.5–8 kilometres] apart, the ships were stopped by a belt of ice. Only this narrow strip between us and letters from home! Those were longing looks which sped across the ice that day." The only way open through the ice led to Foulke's Fiord on the Greenland coast and there Sverdrup found the *Windward*. If not their mail, they were at least given newspapers and told that the steamer they had been so close to had left the mail for them on Pim Island, just off Fram Haven where the expedition had spent the winter.

The first order of business was to fetch the mail. But this Sverdrup just could not manage. He was blocked by the closely packed ice drifting south out of Kane Basin and had to abandon the attempt. This heart-wrenching development was compounded by a more serious consideration affecting the overall success of the expedition. Twice now, *Fram* had failed to force her way through to Kane Basin. Sverdrup's primary goal of exploring the unknown northern and eastern coasts of Greenland had eluded him again. Should he wait and try again the following year or should he put an alternative plan into action? The decision was not easy. Second-best was not in Sverdrup's nature, but then second-best is better than nothing at all—and nothing-at-all was a distinct possibility for the following year as well. Reluctantly, Sverdrup turned his back on Greenland and headed south to explore Jones Sound along the southern coast of Ellesmere Island.

Not being able to achieve his primary goal was a bitter disappointment to Captain Sverdrup. He felt that he was letting his sponsors down. Missing the mail from home rubbed salt into the already painful wound. It was not a happy ship that set sail for Jones Sound on August 22, 1899—none too soon either, for winter was coming on apace and a new home had to be found before freeze-up.

No one had ever penetrated far into Jones Sound. William Baffin had discovered it in 1616, but the Sound had preserved its secrets well. The only explorer to sail any distance into the Sound before the ice stopped him was Commander E.A. Inglefield in 1852. He had gone in some 128 kilometres (80 miles) before turning back. Everything to the west of this point was *terra incognita*. Jones Sound might be second choice, but at least there was good potential for discovery, difficult though it might be.

On his way, Sverdrup ensured his winter's supply of dog food. Tonnes of walrus meat were piled up in a stinking, slimy mess on deck, but conservation was not a problem. Temperatures were already dropping well below the freezing mark every night and barely rising above it during the day.

Jones Sound greeted *Fram* with a combination of impenetrable ice, heavy seas and a violently gusting wind from the east blowing down the funnel between Devon Island and Ellesmere. No wonder the Sound was unexplored. Sverdrup was forced to wait out the storm for four days in a protecting harbour he named Fram Fiord. On August 28, *Fram* stuck her prow back into Jones Sound and met with a world still blowing hard from the east, filled with thick fog, rain, sleet, heavy ice and a strong current to the west. At least the current and wind were in the right direction. *Fram* was held fast in the westward-drifting ice until August 31, when she was finally able to escape and the next day enter the welcome open waters of Harbour Fiord—which was to be her second winter quarters. It was a mere 40 days since leaving Fram Haven, her first winter quarters. The thought of spending yet another winter locked fast in the Arctic ice tugged Sverdrup's mind and heart in the direction of home: "My mind traveled back to Norway, but homesickness is a sure follower in the wake of such thoughts and must not be allowed at any cost."

– 4 –

THE SECOND WINTER

NO MODERN MAN HAD EVER PENETRATED farther west into Jones Sound than *Fram*'s second winter home, Harbour Fiord. The only one to do even that had been Inglefield in 1852. From this point on, Sverdrup's expedition would be exploring new territory.

To prepare the way for the coming sled trips, Sverdrup and four of his men rowed out to the west on September 8 in a small boat full of walrus meat to establish a cache of dog food. They did not leave behind a healthy ship: "Several members of the expedition were unwell, and Peder, usually so cheerful was particularly low. He complained of pains in the chest and spat blood and his legs began to swell..."

A favourable wind blowing in his boat's sail though, also blew away the captain's depressed mood:

> *It was one of the most enjoyable sails I had ever had in my life, combining as it did the desirable conditions of smooth water with a swift breeze behind us. Nor is the element of excitement lacking in this kind of sailing; some dexterity is required to wind in and out among the floes...*

On September 10, the four sailors turned their boat into a finger of water that they named Boat Fiord and camped for the night. Next morning, their craft was immobilized by a heavy jumble of broken ice that had drifted in while they slept. There was a real danger that the fiord might freeze over at any time and the crew had a 70-kilometre (45-mile) row ahead of them before they could sleep on board *Fram* again. Sverdrup decided to go no further. The meat cache would be set up in Boat Fiord, and he and his men carried the walrus meat ashore. The next morning they were ready to leave, but the ice was not ready to let them go. They managed to shoot a seal for food. Hard work in freezing weather develops the appetite. Those developed by Sverdrup and his men can certainly not be classed as dainty:

> *... we hauled it* [the seal] *into the boat without difficulty and filled one of the cooking pots with its blood; we were all very fond of food prepared with this delicacy especially blood pancakes.*

The deeply plunging temperatures of the Arctic autumn soon made it abundantly clear that the foursome could not return to *Fram* by sea, at least not in the boat. They would wait until the frost had strengthened the sea-ice enough that they could walk home, but this might take two to three weeks.

The boat became the roof of their temporary home. First, the men hacked a pit in the dry gravel. Then they turned the boat over on top of the hole and heaped broken rock along the sides, topping the whole structure with a metre (a yard) of insulating snow. The boat was about six metres long and two wide (about 20 feet by 7 feet) on the inside. The depth of the vessel combined with the depth of the hole allowed the men to stand upright under the keel. At the stern was an entrance with a door made of a double layer of sailcloth.

Food during their wait was mainly hares and ptarmigan. They occupied their time by fashioning an Inuit-type sled, lashing together two thwarts from the boat for runners and a couple of harpoon shafts for crosspieces. An empty bread tin was sacrificed to provide material to plate the runners for smooth hauling.

By October 6, the deepening frost had hardened the sea ice enough to support the weight of the men and their sled. They headed home, making it in two days. Late on the second day, as they neared *Fram*, they were met by Bay and Baumann who, worried about their long absence, had set out to find them. The looks on the faces of the two seaman from *Fram* told Sverdrup that all was not well. A terrible thing had happened while he and his three companions were gone. Ove Braskerud was dead.

More than one member of the crew had been feeling unwell when Sverdrup had left a month before. They had been laid low by the effects of the whistling wind, the dampness and the cold they had suffered in Jones Sound before finally taking refuge in Harbour Fiord. Braskerud had taken

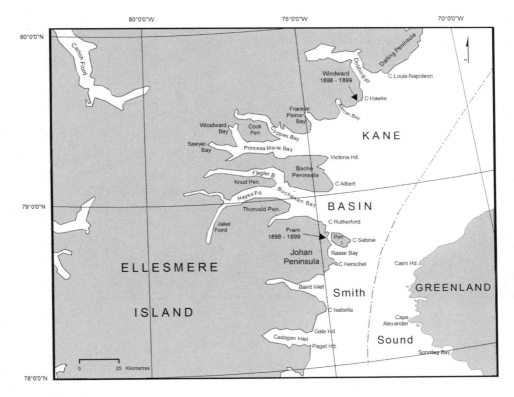

Map 3.

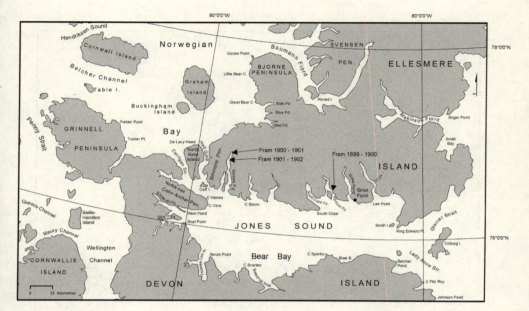

Map 2.

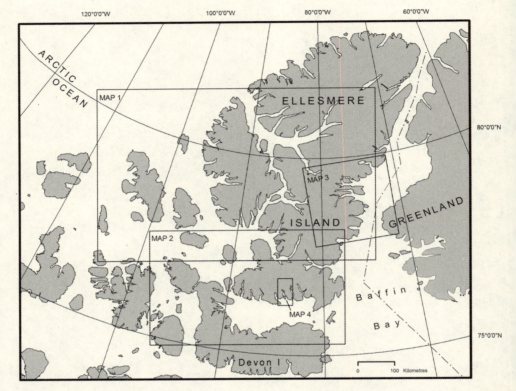

The maps included here represent that area of the Arctic visited by Sverdrup and his crew. The first (base) map shows the entire region; the remaining maps correspond to the inserts on the base map.

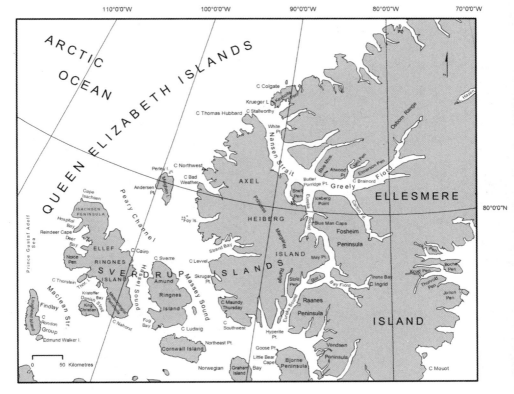

Map 1.

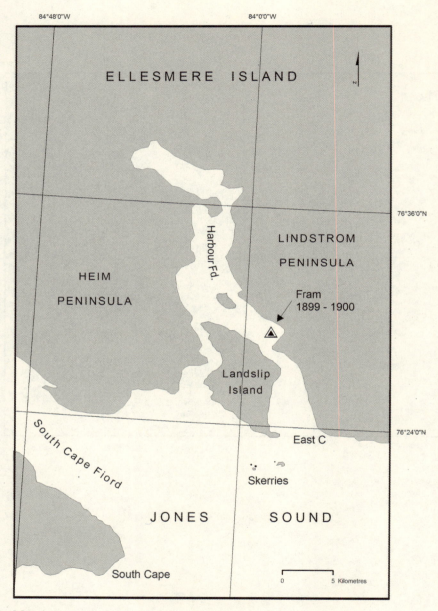

Map 4.

SOURCE: OTTO SVERDRUP'S *NEW LAND*
Ove Braskerud, one of two crew members to die on the Fram's voyage.

to his bed and grown steadily worse. The doctor was now dead and there was no medical help to cure the sick. Braskerud's breathing became laboured and he lapsed into unconsciousness. Finally his breathing stopped.

Once more Sverdrup's mood plumbed the depths of despair:

> Braskerud was a thoroughly good fellow and we were all very fond of him. He had many interests outside his duties and was particularly fond of forestry. In a certain way, his death made a greater impression on many members of the expedition than even that of the doctor, and it caused greater depression among us.

> Winter was ahead, the doctor dead, Peder still ill in his berth, Nödtvedt ailing. And the oncoming polar night, with its cold and crushing darkness, did not tend to help us see things in a brighter light; particularly since the doctor's death, we felt powerless against every sickness.

In spite of death, *Fram*'s crew had to push ahead with preparations for facing another cold and dark Arctic winter. The meat supply for men and dogs had to be secured—walrus for the dogs, muskox, polar bear and small game for the men. As always, the muskoxen were the easiest to kill, twenty animals falling to the crew's bullets in one hunt. In a short time, Sverdrup had what he called a "meatberg" piled up on the deck of his ship—1,700 kilograms (3,750 pounds) of it. He also called the meat his "horrible riches," betraying mixed emotions about having to kill defenseless animals in order to ensure the survival of his crew and himself.

Some 120 kilometres (75 miles) west of *Fram* along the southern coast of Ellesmere Island, at a spot the crew baptised Björneborg (Bearville), Sverdrup set up a cache of polar bear meat to extend the range of the following spring's sled journeys of exploration. Bay volunteered to stay behind

SOURCE: OTTO SVERDRUP'S *NEW LAND*
Ove Braskerud's cross in Harbour Fiord, 1899.

SOURCE: *CANADIAN GEOGRAPHIC*
Photo by B.W. Lidstone of Ove Braskerud's cross, found by the RCMP in Harbour Fiord, c. 1965.

and be the lone guardian of the mountain of bear meat. At least he would not starve!

Hunting polar bear had its exciting moments, both for dogs and men. Once the dogs had scented a bear, nothing could stop them:

> At about two o'clock as we were driving fast up the fjord, a bear came running down toward us. I was determined to get him. We were going downwind, so that the dogs did not get the scent of it at first, but when I suddenly swerved from the course they knew at once that something was going on and increased the pace. They soon caught sight of the bear, and then they set off at such a rate that the snow whined under the runners and the sledge hopped from drift to drift, while I dashed alongside on skis.
>
> I was just about to let go the traces when one of my skis got caught under the sledge and down I went. I still had hold of the connecting lanyard, and it was fastened so tightly to the traces that I could not loosen it. Now it was my turn to be dragged! I can't imagine where the dogs get their strength when they are on to game: the heaviest load is as nothing to them then. Their pace was as fast as ever, though both my skis were now caught crosswise under the sledge.
>
> I did not enjoy my ride one bit. I lay trying to back the dogs so as to unfasten the lanyard, and took one bump after another, each harder than the last. My legs felt as if they would be rubbed right off.
>
> Several times I was on the point of letting the whole thing go, but was afraid the bear would make an end of the dogs, whose movements would be hampered by the heavy load behind them. And my team was so far ahead that it would be quite a while before the other dogs could come to their rescue, so I decided my legs would have to take their chance; they would probably hold this time as they had before.
>
> The bear's position was a first rate one. It had taken its stand on a little plateau high up on a mountain crag; this ledge was reached by a bridge not more than a yard in width. Meanwhile Schei was climbing and scrambling in the snow and stones without seeing the

bear, which was hidden by the ground and did not become visible to him until he was within a few feet, and then it was not long before a shot was heard.

The bear sank to the ground and in a few seconds all the dogs threw themselves onto it. It was very plain that these fellows had not been surfeited with bear-meat. They tugged and pulled at the bear's coat, tearing tufts of hair out of it and before we knew what they were doing they dragged the body to the edge of the plateau, whence it shot out over the precipice.

The dogs stood amazed, gazing down into the depths where the bear was falling swiftly through the air—but not alone, for on it, large as life, were two dogs which had clung so fast to its hair that they now stood planted head to head, and bit themselves still tighter in order to keep their balance. I was breathless as I watched this unexpected journey through the air. The next moment the bear in its perpendicular flight would reach the projecting point of rock, and my poor dogs—it was cruel revenge the bear was taking on them! Now I would only have three dogs left in my team.

The bear's body dashed violently against the rock, turned a somersault out from the mountain-wall, and fell still farther until, after dropping a distance of altogether at least a hundred feet, it reached the slopes by the river and was shot by the impetus right across the river-ice and a good way up the other side.

And the dogs? When the bear crashed against the mountain, they sprang up like rubber balls, described a large curve, and with stiffened legs continued the journey on their own, falling with a resounding thud on the hard-packed snow at the bottom of the valley. But they were on their legs again in a moment and set off as fast as they could go across the river after the bear.

… we put a connecting lanyard through the mighty one's nose and set off. The dogs knew well enough that this meant food for them, and the nearer we came to camp the harder they pulled. In fact, I had to sit on the carcase [sic] to hold them back, and I made my entrance into camp … jolting backwards and forwards *on my new style conveyance. The bear was at once skinned and the dogs had a meal; when they had finished there was hardly any space between their ribs.*

When meat for the winter had been secured, the crew's efforts turned to making ready for the coming spring's sled journeys. The ship turned into a bustling workshop of men, relying of necessity on their native resourcefulness to practice the many trades required for the task at hand. There were no tradesmen to call in. Baumann plied needle and scissors to cut canvas and sew up some two-man traveling tents. Olsen made five new sets of cooking vessels consisting of a biggish cooking pot, a smaller one that went inside, and a coffee pot to go in the smaller one. He also fashioned odometers to measure the distance traveled over the ice and snow. They were made like a bicycle wheel with a steel outer ring surrounding a thin wooden one held to a hub with wire spokes. There were two indicating hands, one reading miles, the other tenths of miles. Olsen was a good tinsmith, coppersmith, instrument maker and gunsmith. Nödtvedt fashioned all manner of iron and steel implements at his forge. When he was not a blacksmith, he tanned skins.

The ship's two salons were turned into workshops. In the after cabin were installed the tinsmiths and the filers of the metal coverings of the sled runners, and hammering, soldering and filing went on the whole day. There too, the broken sleds were mended and the metal plates of the runners put on. In the forward salon, the needle was plied as energetically as the coarser instruments, in the sewing together of skins and the making of clothes, footwear, tents, dog harnesses and sleeping bags. In the 'tween decks Fosheim toiled at his carpenter's bench, and on deck, the clanging blows of the smithy's hammer resounded in the clear Arctic air. Orange and blue flames could be seen flickering through the canvas walls of the forge.

On March 20, 1900, four expeditions pulled away from *Fram*, westbound along the southern coast of Ellesmere. Eight men driving eight teams of straining, howling dogs left on explorations of discovery and mapping of new land.

Two days out, the men stopped at Björneborg to pick up some bear meat. Here they were greeted by Bay, the self-appointed Commandant of Björneborg. The place had got its name from the huge pile of carcasses of polar bears that had been shot and stored there. Bay had suggested himself

SOURCE: OTTO SVERDRUP'S *NEW LAND*
On the trail with an odometer (mounted on the sled), used to measure distance travelled.

as guardian of the treasure lest foxes, wolves and other bears devour it. He then proceeded to spend three months at his post, completely alone at first, then later accompanied by a lone dog as companion, apparently quite happy in his own private microcosm of isolation, within the relatively larger bubble of isolation of the *Fram* expedition as a whole.

On March 23, the four expeditions took their leave of the Commandant of Björneborg, sledding west first past Goose Fiord, then Walrus Fiord and turning north at Calf Island to enter Hell Gate. The strait at this point between Ellesmere Island and North Kent Island got its name from Sverdrup because of the hellish sledding conditions encountered there on the sea ice:

> *To find a passable camping place here was no easy matter, as the ice consisted chiefly of blocks and ice-hills, the latter being at such an angle that it was impossible to pitch a tent on any of them.*
>
> *We began literally to carve our way northwards next morning, accompanied by driving snow and a strong wind from the south. The ice was worse than ever and in places was so impracticable as to baffle description. Towering pressure-ridges which had been*

forced high up against the cliffs in many places compelled us to cut a path foot by foot, through blocks and hummocks of calf-ice. Then suddenly we would be confronted with a fissure so large that the only way of crossing it was to fill it up by shoveling in cartloads of snow. In other places, avalanches had carried boulders, hummocks, and ridges with them into the sea. Where this had happened the snow was as hard as ice, and we had to take to picks and spades once again. The dogs were unharnessed and we dragged the sleds ourselves across the critical spot. High above hung menacing precipices and cliffs supporting enormous cornices which at any moment might fall, sweeping men, dogs and sleds into the whirling stream below.

All this was bad enough, but the weather made things ten times worse. The wind increased steadily, and between two and three o'clock it rose to a violent gale and we could not see our hands in front of us in the driving snow.

With conditions like these, coupled with temperatures ranging down to between -30°C and -40°C (-22° and -40°F), Hell Gate was appropriately named.

The expeditions continued northeast and north along the coast of Ellesmere Island to the tip of Björn Peninsula where the parties split up. Four sleds would continue on—Sverdrup and Fosheim forming one team, Isachsen and Hassel the second. On March 31, the other four men—Baumann, Stolz, Schei and Hendriksen—turned back with instructions from Sverdrup to explore and map areas closer to home.

One very specific task was given to Baumann. He was to search out a land route across the narrow neck of Simmons Peninsula to the northernmost tip of Goose Fiord so that Sverdrup and his companions could avoid having to hack their way though the gates of Hell on the way back. If he found one, he was to leave a note in a rock cairn for Sverdrup to find, showing where the pass was.

A final toast was shared before splitting up:

We had a bottle of brandy for the purpose of drinking to our departure. It was now brought out, and we all presented our cups for our due share of the liquid. The mate, who drew the cork, and was about to pour the drams, suddenly stopped, for nothing came out of the bottle. Could it be empty? Impossible; it was too heavy

> *for that. He shook it and shook it, and tried again; but still nothing came. It was very curious. Then we peeped into the bottle, and found that the contents were frozen, frozen to the very bottom! There seemed to be no good use for our cups, so we got a stick and poked our good "Three Star" brandy out of the bottle. Besides the other object, our parting cup was intended to warm us—there were forty-four degrees below zero [-42°C]—but it was a cold draught.*
>
> *We ate up our dram, said a grave goodby and went our respective ways.*

Sverdrup, Fosheim, Isachsen and Hassel set their sights to the northwest where a large looming black wall of rock beckoned in the misty distance. It appeared to Sverdrup to be some 40 kilometres (25 miles) distant, but before he reached it, he and his crew would travel 110 kilometres (68 miles), and it would take them 10 days, punctuated by a shrieking gale that immobilized them for several days, and shooting polar bears to feed their dogs. When they finally reached the mountain of rock, Sverdrup named it Cape South West. He suspected that the mountain was on a new island never before discovered by modern man, but he could not be sure that it was not just a western extension of Ellesmere Island; he would not be sure until the following year. He named this newly discovered area Axel Heiberg Land—not Island—until he could be certain.

The next night, the foursome set up camp under twin mountain peaks whose suggestive form occasioned some difference of opinion. An unnamed member of the expedition suggested

> *a name more descriptive than tasteful... . Fosheim, who on several occasions had shown himself to be our most gifted advocate of modesty, said nothing, but his face promised ill. ... when in the evening an allusion was made to the same name he declared indignantly that it would not do at all; it was too ugly. No, they would be called The Two Craters, and so they are to this day. Virtue got its reward.*

The unvirtuous name remained undisclosed in Sverdrup's account, but one can imagine what it might have been. On another occasion, modesty did not get in the way of creative geography when Sverdrup named a sinuous finger of sea-water Mück Fiord, which politely translates as Excrement

Fiord. Again one must call upon one's imagination to guess what events might have transpired under the cold Arctic sky that day that the fiord should merit such a name.

Sverdrup and his three companions pushed on from the Two Craters, going north up around the southwest coast of Axel Heiberg Land. They traveled over the frozen highway that was the sea ice. The character of sea ice has many variations. At times it can be quite smooth and blown clear of snow. At others, it can be the worst jumble of jagged, broken-up slabs and chunks tilted at all angles. The following excerpt is from Edward Shackleton's book *Arctic Journeys*, an account of his travels in the region of Ellesmere Island in 1934–35:

> *Axes were in constant use, for the sledges and dog-traces were continually catching up on the ice-pinnacles, and whenever a sledge topped a particularly high ridge it would race down the other side and bury the runners in ice and snow, which we would then have to chop free. But the deep snow that lay in the hollows was the worst feature. Quite suddenly the sledge runners would sink out of sight, the dogs would have to struggle along on their bellies, and we never knew whether the next step would be on a sharp piece of ice, or whether we would find ourselves up to our thighs in snow... . Then, after a few more patches of rough ice, we swung due west on to a wonderfully smooth stretch... . We were now on excellent smooth ice, and we made fast time... . Towards evening, as we neared Hayes Point, patches of rough ice began to occur, and we were finally stopped by a barrier even more impenetrable than the previous day's pressure ice. We had no choice but to force our way through to the mainland about three miles [5 kilometres] distant, in the hopes of finding a good ice-foot... . For the next hour and a half there was such a pounding and jarring of the sledges that it seemed an absolute miracle that they were not smashed beyond repair. It was about midnight, but although the temperature was far below zero, we were all drenched with perspiration, and by the time we reached land our beards, eyebrows, eyelashes and hair were all white with frost, the dogs were completely "all-in." However, we had now reached a broad highway of smooth ice, and we pushed on for a couple of hours, finally camping not far south of Cape Frazer.*

The transition zone from sea ice to land is often an area of intense ice activity due to tidal action and wind. The ice-foot mentioned by Shackleton is a ledge of ice that forms along the shore. It grows out from the shore like a horizontal stalagmitic formation that comes about from the rising and falling action of the tides that continually splash water onto the ever growing ledge. An ice-foot can be up to 30 metres (a hundred feet) or more wide and often makes an excellent sled-driving highway. It can also be quite narrow, and even absent, especially where the land drops off steeply into the sea. The action of the waves and tides can sometimes undercut the icy ledge to the point where collapse is imminent. When the tide is low, the ledge can be 3–3.5 metres (10–12 feet) above the sea ice, so driving along an ice-foot is not without its exciting and dangerous moments.

Danger also comes in other forms when sledding on sea-ice, as Sverdrup and his companion discovered:

> *I noticed several times that there were open holes around the small "hummocks" we passed on the way, but did not pay much attention to them. Also, the sun was shining right in my face, so that it was not easy to keep a lookout ahead. It never occurred to me but that the ice must be strong enough to bear, and I drove confidently on without suspecting any danger.*
>
> *All at once I noticed that the ice was a curious colour. It suddenly became quite a dark blue, and the thin layer of snow on the top of it was wet.*
>
> *We were driving on quite a thin crust of snow and might go through at any moment!*
>
> *This was a cheerful prospect. If we were to fall through here, we would do the thing properly, for it was idle to suppose that we would come up alive from this tearing current—the current which had eaten up the ice from underneath.*
>
> *Instantly, I turned the dogs towards land and tried to keep them to the places where the ice was whitest—we knew it was thicker there—but all the same it was so weak in many places that it sagged under the runners, and sometimes the dogs stepped right through. They, too, quite understood the danger, and I can never remember them obeying the whip as they did that day. In towards land they went as fast as they could go, and Fosheim followed at our heels.*

I do not think that any of us, at any time, was ever so near losing his life as we were then, and I need not say that we were glad when ... we again felt firm ice under us.

Pressure ridges form in the transition zone between sea ice and land. Intense lateral pressures push huge slabs of sea ice against and over each other, and against the immovable shore due to a combination of powerful forces—gale-force winds, rushing currents, the rise and fall of tides and large temperature variations. The ice has only one way to go—up. Huge ridges as high as 10-storey buildings rise up where ice under tremendous pressure from one direction crushes up against ice being forced inexorably in the opposite direction. While passing a mountain of ice on Easter Sunday Sverdrup remarked to Isachsen, "That pressure-ridge is about eighty feet [24 metres] high, I suppose?" Sverdrup admits to understating the height to encourage his companion not to err on the high side. "No," was the response. "it's a hundred and twenty [36.5 metres] if it's one."

It was just such a ridge that Sverdrup climbed on Easter Monday, April 18, 1900, at Cape Levvel (Cape Good-bye) on the western coast of Axel Heiberg Land to see if he could get a clear view over the ice. While he was standing there scanning the country, he became aware of a grayish blue form looming indistinctly, far off on the western horizon. "What could it be? New land? Yes, yes, it was!"

The land he saw was one of the two as-yet-undiscovered Ringnes Islands, most likely Amund Ringnes Island. Sverdrup immediately modified the expedition's course. Isachsen and Hassel would now split off and head west to explore the new land while he and Fosheim continued north along the west coast of Axel Heiberg Land in an attempt to discover if it was an island or just an extension of Ellesmere Island. On their way back, Isachsen and Hassel would also try to settle the question of Axel Heiberg—Land or Island? They would round the southern tip of the land mass and push up into what later became known as Eureka Sound. A number of indications made the expedition leader think that Axel Heiberg Land was actually a new island, as indeed it was, but he could not yet be certain. After splitting up, the two teams of men would not see each other until the following June when they would once more be reunited on board *Fram*.

Howling wind, drifting snow and fog dogged Sverdrup and Fosheim as they struggled north. Contrary to popular opinion, the Arctic is essentially a desert with very little snow cover. A howling wind easily sweeps the landscape bare. The land in this area was extremely flat and sloped away from the frozen ocean at a very shallow angle. The transition zone from sea-ice to land was almost imperceptible, especially with drifting snow and fog to blur the visibility:

> But suddenly to our amazement, we found ourselves completely off course—far inland, in the midst of sandhills and mountains of grit, and all the time we had thought we were driving on sea-ice!
>
> There was only one course for us, which was to go down again; but this was not easy. Wherever we turned, we came upon ridges of grit, across which our runners absolutely refused to slide. We drove round and round in the sand this way for several hours and finally grew so disgusted that we decided to camp. We had another good reason, for Fosheim was very nearly snow-blind and had to retire into the bag as quickly as possible.
>
> The nineteenth of April dawned with a wind from the southeast and drift and fog so thick we could hardly see a thing. In spite of this we found our way through the sandhills back to the sea-ice... We camped in the evening somewhere out on the ice, but whether it was sea-ice or ice above sand we did not know.... Next day, in the same kind of weather, we pressed on as best we could and we had driven about four miles [6.5 kms] when I suddenly discovered that we were standing on the top of a sandhill with a steep drop in front of us. How we got there is impossible to say, for in our innocence we thought we were on level sea-ice.
>
> On the big plains, of which I could just see a glimpse inland, there must be abundant vegetation and, far away from them as I was, I was walking all the time on grass and moss in practically snow-bare country. The moss was so dry and thick that it was like walking on a soft carpet.... We were not long in deciding to keep a course far enough west to be in no danger of again mistaking land for sea.

The weather got worse. For seven whole days, from April 21 to 27, Sverdrup and Fosheim were confined to their tent by the fury of shrieking wind and drifting snow that buried their canvas shelter. Again on April 29–30, they were weather-bound in their tent. For the next five days they struggled northward, but finally had to face the fact that there was no game in the area to feed the dogs and that it would be foolhardy in the extreme to continue. On May 5, the sextant told them their latitude was 81° 55' north, and they built a cairn of rocks to hold a record of their journey that would attest to the fact that they had indeed reached that point.

On May 6, Sverdrup and Fosheim set their sights south into the same poor weather that had been dogging them on their struggle north. On May 16 they reached Cape Levvel, where they had split up from Isachsen and Hassel a month before on their way north. Here they found a short record left by the other party after returning from the Ringnes Islands. Isachsen and Hassel had reached Cape Levvel on April 28, and had continued south the following day.

May 17, Constitution Day, is not passed by lightly by Norwegians. It is the equivalent of July 1 for Canadians or July 4 for Americans. On that day in 1814, Norway gained its freedom from the Danish Kings and proclaimed the constitution which is still in force today. Every year, it is a day of great celebration in Norway, and Sverdrup and Fosheim were not to be denied:

> *Our Seventeenth of May programme had long occupied us in the evenings inside the tent, for although we were on a sled journey we intended to celebrate it as well as our circumstances would allow. Our breakfast at Cape Levvel, which was not to be despised, consisted of biscuit fried in butter and fat, to which we added "buttered eggs." The festal dinner, later in the day, consisted of a kind of daenge, specially* [sic] *invented in honour of the day, and which in addition to the invariable percentage of water contained the following delicacies: pounded buns with sultanas, figs, nectarines, chocolate, egg-powder, and butter. The compound tasted almost divine, and no housewife will regret including it as a dessert in the most recherché menu. When we had eaten daenge until we could eat no more, we had coffee and the last remains of our brandy. Fosheim contributed two cigars as a surprise, having brought them with him for this express purpose. Finally came the great commemorative event of the day, in the shape of a song composed in honour of 17 May in latitude 79° north, at Cape Levvel. It was sung by Fosheim solo, with the howling of the dogs by way of chorus, for no sooner had he reached the third verse, at which point Gulen* [a sled dog] *and his brave choir were called upon as auxiliaries, than the dogs, who must have thought it a curious proceeding, began to howl and whine in the most doleful*

manner.... It is not easy to explain the reason, but the fact remains that they went on howling to the bitter end.

Perhaps it was just as well that there were no witnesses to this scene of Arctic madness, for it might have been surmised that the two explorers had become infected with a serious case of bush fever and had taken leave of their senses. In spite of their celebrations and any possible after-effects, Sverdrup and Fosheim carried on south the next day, for they found themselves "very fit and more inclined for work, perhaps, than many at home in Norway on the day after the Seventeenth of May."

As the expedition neared Fourth Camp on Ellesmere Island at the entrance to Hell Gate, a polar bear had the misfortune of being in the vicinity. There was no holding back the dogs, which were on the verge of starvation. Their job was to harass the bear and make him stop until the men caught up with their guns. The dogs were so hungry that they might have finished the job themselves, though not without serious casualties. Fosheim killed the bear with two shots, at which the two teams lay down on the ice, one on each side of the bear, keeping watch:

> *In spite of their ravenous hunger, they waited patiently until it was skinned—they knew it was not their turn till then. But I never saw anything like the way they ate when the time came for them, and I have been witness to a good many scenes of the kind. When we had spread the skin on the ice and pitched the tent on it we cut some meat in thick slices, and fed the dogs as long as they were able to swallow.*
>
> *For ourselves, we fried some delicious steaks. It was one of the most beautiful evenings we had on the entire journey. The temperature had risen to an astonishing extent since we had come down here. Up in the bay it had varied from zero to -4° Fahrenheit [-17.7° to -20° C], but now we had 28° [-2.2°C]. The evening was so calm and peaceful, and the sun so warm...*

At Land's End the next day Sverdrup and Fosheim found a cairn which Baumann and Raanes had put up earlier in the month. Sverdrup had instructed Baumann to find a land route across the neck of Simmons Peninsula into Goose Fiord to avoid the terrible sledding conditions encountered in Hell Gate on the way north two months before. In the cairn was a note and a sketch-map from Baumann describing a route from Reindeer Bay to the isthmus separating Goose Fiord from Walrus Fiord.

The next morning as they started across the neck of Simmons Peninsula, the pair was greeted by thick fog and heavy snow,

> ... and so it happened that we found a rather different way overland into Goose Fiord than the one Baumann had indicated.
>
> After passing over the watershed we came down into a pretty [river] valley, and this we followed, well pleased with ourselves. But by degrees, as we descended, the valley became narrower and narrower, and we began to wonder whether, after all, it was going to end in a canyon. Without any warning we were suddenly stopped by a high wall of ice which entirely cut off the valley. We made a halt to see if we could find any way of advance, to avoid driving the long distance back again, but the ice was absolutely perpendicular and inaccessible to any being without wings.
>
> Suddenly it occurred to me that somewhere or other the river must have an outlet. There might perhaps be a tunnel through which we could pass, and on looking behind a massive snow drift I actually found a big hole which, on investigation, proved to be the beginning of a very large tunnel which pierced the glacier. A journey through it, however, did not seem very alluring. From the roof were suspended big blocks of ice which might fall at any moment, and indeed a good many had already done so during the course of the winter, for on the polished river-ice lay masses of blocks, large and small, which pointed an unequivocal warning to the danger of passing that way.
>
> I confess I did not feel very much tempted by the idea of going through it. I went back to Fosheim and told him what I had seen, adding that we need not decide before we had had something to eat.
>
> Not a word was exchanged while we were at work—but for that matter we did not as a rule talk much at meals; we had enough to do without it. When we had finished, and had lighted our pipes, I decided we were capable of making a decision and so asked Fosheim what he thought we ought to do.
>
> He waited for a while before he answered, then took his pipe from the corner of his mouth and said: "We must try. It would be the very devil to drive all that way back."

"Very well," I answered, "but it must be each man for himself. It would be terrible if a block of ice fell on the heads of both of us. At least one must get through."

I shall not forget the moment when we entered the tunnel. I did not feel brave—I openly confess it. In fact I was afraid rather than otherwise. And yet it was not fear that had the hold on me, but rather an uneasy feeling of awe.

Here were lofty vaults and great spaces between the walls. From the roof hung threateningly above our heads gigantic blocks of ice, seamed and cleft and glittering sinisterly. And all around were icicles like steel-bright spears and lances, piercing downwards upon us. Along the walls were grotto after grotto, vault after vault, with pillars and capitals in rows, like giants in rank, and over the whole shone a ghostlike bluish-white light which became deeper and gloomier as we went on.

It was like a fairyland, beautiful and fear-inspiring at the same time. I dared not speak. It seemed to me that in doing so I would be committing an act of desecration. I felt like one who has imperiously broken into something sacred which Nature had wished to keep closed to every mortal eye. I felt mean and contemptible as I drove through all this purity.

The sleds jolted from block to block, awakening thunderous echoes as they went. It seemed as if all the spirits of the ice had been aroused and called to arms against the invaders of their church-like peace.

I breathed more freely when I saw a glimpse of daylight in the distance, and so no doubt did Fosheim. We looked at one another. It is very wonderful, now and again, to come right under the mighty hand of Nature.

From the tunnel, down the remaining part of the valley, we drove without hindrance of any kind. When we came out on the fjord we discovered that we had inadvertently laid our course a good way south of the isthmus between Goose Fjord and the head of Walrus Fjord. We therefore took a line south-east across Walrus Fjord to the outer isthmus, where we camped.

The next day Sverdrup and Fosheim set off for Björneborg, arriving at 10:30 a.m., finding Bay, the so-called Commandant of Björneborg, still asleep in bed. The inner fortitude and self-sufficiency of these hardy Norwegian explorers was remarkable. Bay had spent three months, with only a dog as companion, guarding this depot of meat against the ravages of scavenging animals. Fosheim, who had spent 73 days on the trail with Sverdrup and was now only three days away from *Fram*, was not yet to rejoin his companions on board ship, for he now took over as Commandant of Björneborg, thereby releasing Bay from his lonely vigil.

On June 3, 76 days after leaving, Sverdrup arrived back on board his ship, to be greeted by the news that *Fram* had just barely escaped the most potentially disastrous calamity possible for an Arctic expedition—destruction of their vessel by fire!

Fram was a virtual time-bomb awaiting the right combination of circumstances to burn herself to ashes and sentence her crew to almost certain death in the unforgiving Arctic. A canvas awning had been set up on deck for the winter. Under the awning were stored 16 kayaks, but not just ordinary kayaks. These had been treated with paraffin to waterproof them—paraffin is what candles are made of because it burns so readily. This was the fuse. Near the kayaks was an iron tank filled with alcohol. This was the priming charge. And next to the tank stood the ship's supply of gunpowder. This was the bomb that could blow *Fram* into a thousand pieces and scatter them over the hard, frozen surface of Harbour Fiord, effectively writing a new chapter in the book of the disappearances of Arctic expeditions.

On Sunday, May 27, 1900, at about noon, a spark from the chimney of the galley fell unnoticed onto the canvas awning. Simmons, who was walking on deck, was the first to discover the blazing awning. He gave the alarm and seconds later the crew of nine men was on deck frantically trying to contain the flames. Above the awning, the mainsail was afire and below it the flames had eaten their way into a pile of dry thin boards. Soon the 16 paraffin-impregnated kayaks were engulfed by the flames which licked their way up the mainmast, setting it on fire and destroying the winding tackle on both sides. The men worked like demons dragging the boxes of gunpowder one by one away from the sea of flames. The deck grew burning hot as did the tank of alcohol which was placed in such a way that it could not be moved. The tinning on the outside of the iron tank melted, but still it did not explode. If it had, the ship's fate would have been sealed. The fire spread to the 'tween decks. But the men were far from idle:

> *Luckily there was plenty of water to be had close by the ship's side, and when the cases of gunpowder had been safely removed, the work of extinguishing the flames went on rapidly. Bucket and bucket of water was thrown, hissing and steaming over the deck; men shouted and ran in and out of the heat, the sweat on their brows and their hands black with soot. They hurried backwards and forwards to the ice and up to the vessel again; for our craft was dear to us, and fight they meant to for every plank of the old Fram—our only bit of Norway up there in all that solitude.*

Fram was saved but not without considerable loss—not so much to the ship itself as to the kayaks, all of which were destroyed, their skis, a quantity of wood, muskox and polar bear skins, sails, rigging and sailing tackle of various sorts. The deck and mast were only slightly burned. The hull was not damaged to any extent. The ship carried plenty of sailcloth and rope below decks, so her rigging was soon put in order again. The men realized how lucky they were to have come out of such a potentially devastating event with so little serious damage.

On June 19, Isachsen and Hassel, the members of the other major exploration team which had split from Sverdrup and Fosheim at Cape Levvel more than two months previously, arrived back on *Fram*. In keeping with Sverdrup's instructions, they had pushed north into the wide fiord—or sound, they knew not which yet—that is known today as Eureka Sound. At that time, it was not known if the body of water was a fiord that came to an end at its far extremity, or a sound that connects up with a large body of water at both ends, which would have confirmed Axel Heiberg Land as an island instead of an extension of Ellesmere Island. It had been Isachsen's task to find out.

In fact, Isachsen mistakenly concluded that the wide body of water was a cul-de-sac fiord, ending abruptly at Stor Island. He failed to see the channels leading north on both sides of the island. This error was to add considerably to the expedition's labours the following year, but Isachsen must not be judged too harshly. Visibility was so poor that he and his partner had trouble distinguishing land from sea ice. Although he penetrated the entrance of Bay Fiord, a week of almost continuous snow prevented him from taking position sightings. He did not recognize this body of water as being the same one discovered by Sverdrup and Bay the previous year on their east-west crossing of Ellesmere Island from Fram Haven.

During the extended absence of the two major dog-sled expeditions manned by Sverdrup-Fosheim and Isachsen-Hassel, other members of *Fram*'s crew made three additional important expeditions. In all, these five expeditions spent 282 days on the trail and covered 8,840 kilometres (5,500 miles)—a major piece of exploration and discovery of new land never before seen by modern man.

The thaw came quickly to Harbour Fiord in the summer of 1900. By July 19, the powerful energy of continuous and unsetting sun set *Fram* free from the large floe that had held her prisoner throughout the fall, winter, spring and early summer. But this did not mean that the ship was free to go. Not yet. Although the fiord was soon completely clear of ice, there was still a solid icy barrier across its mouth that prevented Sverdrup from moving out into Jones Sound. It was

SOURCE: OTTO SVERDRUP'S *NEW LAND*
Gunerius Ingvald Isachsen (left) and Sverre Hassel (right), back from a long dogsled expedition.

not until August 9 that *Fram* could steam out of her second winter quarters, ram her way through a day's worth of ice in Jones Sound and head west over a sea that was free of ice and smooth as a mirror.

Sverdrup's plan was to sail as far west as he could before being forced into winter quarters for yet a third year. The further west he could sail, the better would be his position for starting dog-sled expeditions the following spring to complete the exploration of the new lands he and his crew had discovered. In the days that followed, *Fram* sailed through relatively ice-free water along the southern coast of Ellesmere Island as far as Colin Archer Peninsula, then northwest up through Cardigan Strait and as far along the coast of Devon Island as the north side of Arthur Fiord. From this point, ice became a problem. On August 14, the expedition not only was brought to a grinding halt by the ice, but in fact was inexorably forced back in a southeasterly direction by its drift, back to the northern end of Cardigan Strait. There was nothing for it but to submit to the irresistible force of the icy drift down Cardigan Strait and to try to work the ship around the southern tip of North Kent Island and up through Hell Gate. But the ice gods would have none of it, for *Fram* next found herself being forcibly blown back up through Cardigan Strait and into Belcher Channel where she became stuck fast in the closely packed ice.

The outlook was not bright. There was every possibility, indeed, every probability, that the expedition's winter quarters would end up being in a very exposed and undesirable position, beset in the middle of the frozen sea, far from the protective shield of a snug, mountain-surrounded harbour. On September 3, Sverdrup accepted the inevitable. He put out the fires under the ship's boiler and emptied it of water. The engine was secured for the winter. Preparations for a third winter had begun. By September 13, blocks of ice had been sawn out of the frozen sea and a smithy built out of them on the sea-ice. A roof of wood was put on top and the forge was moved in. By September 11, Sverdrup deemed it safe enough to let Baumann and Raanes go off on a short shooting expedition in the direction of Arthur Strait.

On September 14, there began a series of events that could not have been foreseen, nor in all objectivity, even hoped for. Sverdrup noticed a large lane of ice-free water opening up a few hundred metres (yards) to the west of the ship, but *Fram* remained as solidly frozen in as ever. By the following morning, though, the ice was beginning to move, and free water was beginning to form around the ship's hull. In less time than the captain would have thought possible, the ice surrounding his vessel broke up into rubble under the force of tides, currents, wind—he knew not what. To his great surprise and joy, his ship was floating free of its icy shackles.

There was no time to lose. The forge and all the smithy's tools were hoisted back on board. The smithy itself drifted away together with its valuable wooden roof.

For two days, *Fram* drifted where the ice took her, powerless to do otherwise in the tightly packed jumble of floes. The expedition was dragged along by the drifting ice to a point just west of Graham Island. But then, on September 16, Sverdrup noticed that the ice to the south was getting somewhat slack. Here was a stroke of luck! Orders were immediately given to fill the boiler, light the fires and get up steam as quickly as possible. At this point, the men caught sight of their drifting smithy and were sent off to recover the wooden roof.

Sverdrup forced his ship through the ice and was soon steaming toward land to the south in ice-free water.

Baumann and Raanes were still out on their shooting excursion, and this worried Sverdrup. If they were not found, a cache would be built on shore with supplies and a note telling them where the ship was headed. Men with sleds and dog-teams would be sent out to pick them up as soon as possible. Fortunately, as the vessel steamed south, a lookout in the crow's nest spotted the pair driving their sled along the shore of Grinnell Peninsula on Devon Island. Sverdrup set his course for a bay where he knew there would be a shelf of good solid ice frozen to the shore. Baumann and Raanes saw what their captain was about and sledded over to meet him. The ship nestled up to the ice-shelf; crew members lowered ropes over the edge and hauled up sleds, dogs and the two men, and in the next minute *Fram* and her crew were steering for Cardigan Strait, unexpectedly free of their prison of ice and hopefully searching for a more snug winter harbour.

Sverdrup headed south through Cardigan Strait through easy sailing waters and twice tried to steam north through Hell Gate, meeting solidly packed ice both times. The only choice left was to sail east and up to the north end of Goose Fiord where Sverdrup anchored his ship on September 17, 2.5 kilometres (1.6 miles) from the head of the fiord. This would be far better winter quarters than the unprotected, gale-swept expanse of Belcher Channel where *Fram* had been frozen in solid two weeks before:

> *Everything in Goose Fiord was beautiful to our eyes, and a better winter harbour we could not have wished for. The storm which was raging outside had not reached in here. The fjord was free of ice, the land bare, the air mild. It was as if we had come to an Eden.*

– 5 –

THE THIRD WINTER

According to plan, the *Fram* expedition entered its third and last season of Arctic exploration. Sverdrup and his remaining 13 crew members began looking forward to sailing home at the breakup of the ice late the following summer. But first, the crew embarked upon the now almost familiar round of preparations for ensuring survival in the deep cold of the coming winter in high Arctic latitudes. A successful autumn kill once again guaranteed an ample supply of meat for men and dogs. A forge built of ice blocks was set up on the sea ice next to the ship. The blacksmiths, carpenters, tinsmiths, tailors, tanners, shoemakers, tentmakers, butchers and cooks again plied their many trades preparing the equipment and supplies that would be needed for the major spring sled journeys of exploration.

The expedition had barely settled into Goose Fiord when Olsen, *Fram*'s engineer, had good reason to regret the death of Dr. Svendsen, but as the old saw goes, necessity is the mother of invention.

On October 18, Sverdrup and Olsen set out on a 10-day sled trip. A howling gust of wind caught Olsen's sled and blew it across the ice like a leaf, crashing it into a block of ice. Olsen shot off his sled and crunched down hard on his shoulder, dislocating his arm. After a few days of most painful traveling, Olsen and Sverdrup arrived back at *Fram*:

> As soon as we arrived on board I set Simmons to find some of
> the doctor's books and see what we had better do for Olsen's
> arm. We found some diagrams and various directions as to how
> a dislocation should be reduced, and after some consideration
> chose the way which seemed easiest and most simple.

The operation would have been easy enough had we dared to chloroform our patient, but we had no desire to attempt such a thing. What were we to do? Several days had elapsed, and the arm was swollen and angry. Inexperienced as we were we should probably torture poor Olsen most horribly before we got his arm into place again.

I therefore decided to make him thoroughly drunk—the effects of that we could better grapple with. For this purpose we first tried naphtha [white alcohol?], but that did not do; he disliked the taste of it so much that I could not bring myself to force more on him. Good—we had other things that tasted considerably better. I entered into a partnership with the brandy fiend; sent for a bottle of the best Cognac; and began to give him dram after dram. But it was really too much to expect him to drink himself half-seas-over on dry nips all alone, without any other diversion, so I sat down and talked to him about everything I could think of. At first he was very much taken up with his arm, but we went on to the expedition in general, then to shooting in general, and lastly, after innumerable excursions, landed in the Lofoden Islands, in which as a Nordlaending he was much interested, and had himself taken part in the fisheries there.

In this way I brought him little by little into brilliant spirits; he grew livelier at every dram. Fosheim and Simmons, who had been chosen for the deed of bone-setting, sat awaiting the propitious moment, following with much excitement his various "stages of development" during our potations, while I talked myself blue in the face to get him to drink more, and hasten on the crisis of this tragi-comedy.

It was not long before Olsen himself began to be highly pleased at the whole performance, declaring it was the most amusing entertainment he had ever taken part in!

When he had swallowed about a half a bottle of brandy we thought he must be about ripe to be taken. We accordingly placed him on a chest, and the bone setters began their work; but no—the arm would not go in. In his semi-conscious condition Olsen took the whole thing with the greatest calm and said nothing when Fosheim and I then tried our hand on him. To our surprise we were successful at the first attempt! That it was with unspeakable relief we heard the crack of the arm as it slipped into its socket, I need hardly say. As for Olsen, notwithstanding all he had taken down, it had not had much effect on him while we were doing our worst—the pain and excitement had kept him sober—but the instant the arm was in its socket he became dead drunk.

He was carried to his cabin in a hurry, put to bed and a man set to guard him. We thought perhaps he might become delirious, or something of the kind, for, as I said before, he was not quite sober. Fosheim took first watch. But Olsen behaved nicely the whole night, and next morning was quite himself again. We bandaged his arm so that he could not move the joint, and thus he was to go for three or four weeks.

Olsen's happiness, after he went about with his arm in the sling was quite touching to behold. He had not had the slightest hope about himself, and during the agony he went through had painted the future in very gloomy colours. If Olsen was glad, we quacks were no less so, and proud into the bargain. We had discovered a brand-new Arctic surgical treatment, with the brandy fiend himself as assistant. But it is ever the same: genius is simplicity, and evil for evil is only fair play.

The following spring, Sverdrup's scientifically inquisitive mind would not rest until he had settled two important geographical questions: was the area he called Axel Heiberg an extension of Ellesmere Island, or a new island on its own; and what was the extent of the new land briefly visited far to the west by Isachsen and Hassel the previous year? In both cases, the coming expeditions would carefully map large areas of new land that the modern world did not yet know existed.

Open water in Norwegian Bay the previous autumn had prevented the establishment of advance food and supply depots along the planned spring exploration routes. This had to be done in late winter. On March 12, two parties sledded away from the ship loaded down with gear, cutting across the neck of land separating the northern extremity of Goose Fiord from Nordstrand. The two parties spent a -50°C (-58°F) night together at Great

Lake before splitting up the next morning. Isachsen, Hassel, Raanes and Schei headed west to set up a depot at Cape South-west on Axel Heiberg Land to support the coming expedition to the new lands in the west. Sverdrup, Fosheim, Hendriksen and Baumann would reconnoitre Baumann Fiord east of Björne Peninsula and establish a depot there. What Sverdrup did not know at the time was that his decision to explore in the Baumann Fiord area was based on the erroneous conclusion drawn by Isachsen the previous year that Eureka Sound was not a sound at all, but merely a fiord closed off at its northern extremity. Sverdrup was still convinced (correctly so, as it turned out) that Axel Heiberg was an island and that there must be a sound somewhere between Ellesmere and Axel Heiberg connecting Norwegian Bay to the open sea further to the north. Sverdrup would search for this sound everywhere but where he should have, that is, Eureka Sound. Isachsen's error was to cost the expedition many frustrations and many extra kilometres of tough sledding. March 24 saw the depot-laying expeditions back on board ship having spent 13 days on the trail during which the temperature averaged -45°C (-49°F).

On Easter Monday, April 8, six men left *Fram* on the major spring expeditions, each with his own team of dogs straining every muscle to pull a heavily loaded sled over the sea ice. Isachsen and Hassel were returning to the land they had discovered to the west of Cape South-west the previous year in order to map it and, if possible, determine its extent. Two parties of two men each, Sverdrup-Schei and Fosheim-Raanes would try to solve the riddle of the missing sound separating Ellesmere from Axel Heiberg.

On April 9, the parties split up, Isachsen and Hassel heading northwest toward the Ringnes Islands and the other two parties northeast toward Baumann Fiord, where they found their food and supplies depot four days later. Sverdrup was not in the habit of complaining about harsh conditions in his writings during the *Fram* expedition. The fact that he did express a complaint on this day, be it ever so mild, certainly indicates that conditions were atrocious:

> *We reached our depot in Baumann Fjord on 13 April and spent the next day digging out the supplies we had left there and loading up for the journey north. None of us found this delving in the snow very agreeable, to say nothing of taking observations and the like with the temperature at -56°F [-49°C].*

At this point began the search for the missing sound that Sverdrup was almost certain would confirm Axel Heiberg as an island. He first targeted the body of frozen water that he later named Baumann Fiord as the most likely candidate. But he was wrong. It ended up in a narrow, north-tending appendix which he appropriately named Vendom Fiord—Turnback Fiord in English.

At the junction of Baumann and Vendom Fiords, the expedition discovered an intriguing geological feature—coal. Coal was formed many millions of years ago when thick, heavily matted vegetation of warm and humid land gradually sank and was subjected to extreme pressure and high temperatures over many eons of time. Ellesmere hardly fits the picture today. One possible conclusion is that the Arctic lands were once in much warmer, perhaps even tropical, latitudes that eventually drifted to the north. Geologists, however, claim that this is not what happened. For some unexplained reason, the overall temperature of the earth appears to have been much warmer eons ago, causing a moderate climate even at high Arctic latitudes. Years later, in 1985, another fascinating indication of warm temperatures in the Arctic in prehistoric times was found on Axel Heiberg Island—tree stumps. These stumps are not petrified. They are still wood, preserved in the aseptic conditions of the frozen Arctic. Even after 45 million years, the wood can be carved and it still burns.

Sverdrup and his three companions next sledded north into Troll Fiord, but were again disappointed to find that the quickly narrowing finger of ice soon came to an abrupt end at "an ugly place shut in on both sides by high and gloomy walls of rock."

In spite of his lugubrious surroundings, Sverdrup still did not give up hope, but hunted out some possible valley at the end of the fiord that might lead them to the north. A fairly wide valley opened up on the east side of the fiord which the expedition followed up onto a high plain where they set up camp.

Game was abundant. A couple of muskoxen were shot for food. The next day Sverdrup scared up another herd:

> *As they started to run away I noticed that one of them had a new-born calf. The herd went up a steep snowdrift, eight or ten feet* [2.5–3 metres] *in height, and the calf made a brave attempt to follow, but when it had almost reached the top it lost its footing and rolled down to the bottom again. It fell so violently and helplessly that I thought it would be killed, but to my surprise it rose to its feet and began to scramble up once more. Its*

> second attempt to scale the drift was no more successful than the first, and again it came rolling down. It cried piteously, like a baby. I felt so sorry for it that I was just starting to help it up the drift when it occurred to me that the old cow might misinterpret my motives, and what then? I might risk a battle with her, and it would be a pity to have to shoot her in self-defense. I decided to stay where I was and await the turn of events.
>
> At last the mother heard the cries of distress and came tearing down the hillside, the snow flying behind her. Heaven help the person who had meddled with her calf then! She would have made it hot for him. The mother caressed the calf as if to comfort it, sniffed it all over to see if it was hurt, gave it a push now and again, and then started gently up the drift; but not the way the calf had tried to go in following the herd; she carefully chose an easier and less steep path.
>
> When she had got the calf across the drift she ran a few steps forward, not very fast, but too quickly at any rate for the calf to follow her. Then she turned back and pushed it from behind with her muzzle, so that it went a little faster. Again she ran a few yards forward, but still the poor little thing could not keep up with her, and she went back to her old pushing methods.
>
> And so they went on, all the way up until they reached the square [the protective formation of the animals]. Then she took her place in it, and the calf crept under her and was entirely hidden from sight by her long hair.

Big game was the mainstay of the expedition's meat supply, but smaller animals provided welcome variation from muskox and polar bear meat. Ptarmigan, ducks and Arctic hare were not passed up. When not shooting the big Arctic hares, Sverdrup found them amusing creatures. The plateau on which he was camped and the canyon leading down from it were overrun with them:

> On the way down I saw a score or so of hares sitting nibbling the grass on a little hill. I made my way very slowly towards them, just to see how near they would let me come. They soon caught sight of me, and slowly collected. Before long they were an unbroken white mass, with their heads inwards and their tails out.

There were so many of them that there were several rings, one inside the other, and it was a life-and-death matter to be in the innermost ring—at least, so it appeared to me, for they made the greatest commotion about it. They pushed and fought and bit each other until they screamed aloud, all the time slowly revolving, something like a millstone.

So this was the square of the Arctic hares!

After watching them for a while at ten or twelve paces I moved away slowly in order not to frighten them right out of their wits. After I had gone a little way off I saw on looking back that they had begun to disperse and were browsing again.

Another time,

I was soon able to count thirty-one animals. Thirty sat motionless the whole time, looking as if they were asleep, but the thirty-first was plainly a sentinel. She hopped about, in and out among them in never ceasing vigilance. Every now and then she sat down and listened for a time, but not hearing anything to arouse her suspicions, continued her rounds among the sleepers.

I made my way towards them with all the stealth I was capable of, but it was not many minutes before the sentinel noticed me and became upset. Every time she showed signs of alarm I stood still for a while, and when her fears were allayed took another step or two forward. But no sooner did I begin to move again than she scrutinized me as sharply as before, and again grew frightened.

I had plenty of time, and took things quietly, so that in the end I got within quite a short distance of the hares. But at the last moment the sentinel apparently thought me a little too forward and suddenly started to run frantically around her flock, striking the ground with her hind-legs till it quite resounded. Then she set off up the slopes with all the others after her in a long, straight line, looking as if a white cord had been stretched up the hillside and over the ridge at the top. I continued to look for a time after they had disappeared from sight. The whole thing was so strange that I wanted to think it out.

Not far from me still sat two hares, by themselves. Evidently they did not belong to the other lot. I thought it would be interesting to go across to them if possible, and see what they were about, but realized that I would have to make use of other tactics if I wanted to get near them. This, I thought, would be a fitting moment to impersonate a reindeer, or some other kind of big game, and I made a valiant attempt to simulate their grazing movements, backwards and forwards on the grass. Meanwhile I kept a sharp eye on the hares, and always took care to move a little nearer to them.

The hares soon noticed the ever-advancing figure. They stood up on their hind-legs and started gazing at me. I immediately stopped, remained quite still and gazed back at them. When they were quite reassured, I began to move about on the grass again, and at last they grew so accustomed to my presence that they did not take the slightest notice of me.

My tactics were so successful that, in the end, I was not much more than two or three yards [2–3 metres] from them. It was quite touching to see the great, innocent Arctic hares sitting only a few paces off, quietly gnawing roots. The only attention they gave me was an occasional sniff in my direction.

As I stood watching them, one of the hares came quietly up towards me. So near did it come that I stretched out my hand to stroke it, but this it did not quite like. It started a couple of paces aside, and then began quietly to eat again.

For a long time I stayed fraternizing with the hares down on the grass, and at last we did not mind each other in the very least. They went on with their occupations quite unconcernedly; and I with mine. I felt like Adam in Paradise before Eve came and all that about the serpent happened.

On April 28, all four sleds slipped down off the plateau into a canyon that fed into a large wild valley running in a northwest direction. Herd after herd of muskoxen lifted their shaggy heads from grazing on the vegetation poking through the sparse cover of snow and stared with uncurious eyes at the strange procession of creatures sliding past. By evening, Sverdrup could see that the valley was leading down to a large fiord just a few kilometres in the distance. Excitement was mounting among the explorers. They felt that they were near to finally discovering the solution to the riddle of Axel Heiberg—island or land? The next morning, Sverdrup went scouting down to the shore of the east-west running fiord. What he did not know was that this was the very same fiord that he and Bay had first seen two years previously when they had crossed Ellesmere Island from their first winter quarters at Fram Haven; he had named it Bay Fiord. He also did not know that it was the same fiord that Isachsen had peered into through fog and heavy snow the previous year when he had mistakenly decided that Stor Island was the northern limit of his so-called Great Fiord. In certain ways, the lay of the land did not resemble anything reported on by Isachsen. On the other hand, the Gretha Islands did jibe with his findings. Sverdrup was puzzled. "The whole thing seemed very mysterious," he mused in his writings, but he felt that the mystery was on the point of being dispelled:

> We drove up to the crack, made a short halt, and fortified ourselves with bread and meat; for it was necessary to be equal to the situation.
>
> We were about to come face to face with a solution—that we all sensed. But whether it would bring us bitter disappointment or jubilation was still a closed book.
>
> We then went a little way up the talus. Never before had we scanned land and shore with such excitement.
>
> And what did we see? A beautiful large sound extending northward as far as the eye could reach!
>
> We were looking into the promised land and we were as happy as children!

Sverdrup now knew that Isachsen had been tricked by poor visibility the year before. Oh! what extra labour and lost time this error had caused them as they had futilely chased the wild goose first up Vendom Fiord and then up Troll Fiord. Isachsen's Great Fiord was renamed Eureka Sound by Fosheim. At last they knew. Axel Heiberg *Island* it was!

"Out with the bottle!" The four explorers toasted their hard-earned success with Norwegian schnapps. "If we had failed to reach this point, I think I would have considered the whole spring season wasted," Sverdrup wrote.

The expedition pushed north into newly discovered Eureka Sound. On May 4, it split up into two teams—Sverdrup and Schei followed the western shore of the fiord, reaching as far north as Butter Porridge Point on Schei Peninsula before calling it quits for the season and heading back to *Fram*; Fosheim and Rannes explored the eastern shore as far as Greely Fiord and down into Canyon Fiord before turning around. The two teams of explorers returned home independently of each other.

May 17, Constitution Day, once again found Sverdrup and his partner on the trail. The usual celebration was not to be denied:

> *For several days all our spare time had been devoted to planning how we would observe the Seventeenth of May. Olsen had presented Schei with a tin of fruit, and the problem was how to make the most of it. After racking our brains for some time we settled on a pancake, the fruit to be eaten with it. It was the first time we had attempted a pancake on a sled journey, and it would not be strictly correct to say that we did it then. The ingredients were all right—egg-powder, flour, and sugar—but we could not get them to bind, and when the pancake was ready it looked more like a kind of thick porridge than anything else. It tasted excellent, however, and that was the main thing. We had no brandy, but quenched our thirst with strong coffee, and our spirits rose to great heights.*

But the Constitution Day excitement was far from over:

> *At about eleven, we were just going off to sleep when the dogs began to give tongue. "Ha, ha! There's a bear," we thought and made for the door. But we could see nothing. On the other hand, Sergeanten and Svartflekken seemed to have been blown away. What could this mean? I began to call them and at last they came running up. We tied them both up again and crept back to bed, still wondering why they had broken loose and run away.*
>
> *We talked a while, but little by little the conversation flagged, and we had just gone to sleep when an uproar outside brought us both to our feet. The dogs were yelping and howling as if the end of the world had come. Schei was the first to stumble out through the door, and I was at his heels. Just as I was creeping out I heard him say, "What the devil is this?" and saw him seize his gun. I was quickly out too, and had hold of my gun.*

A large pack of wolves was making an end of my team! I had been obliged to tie them up in two lots on account of the three new dogs, which had been quarreling at night with the older ones. As we came out six or seven wolves were trying to tear one of my old dogs to pieces. They had broken his rope and dragged him a hundred yards [90 metres] *from the camp. Five or six more were standing ready to begin on my three new dogs.*

Schei immediately opened fire, and I followed his example as soon as I could. The minute the firing began the pack let go the dog they had under them and made off. I felt the poor thing was long since dead and was not a little surprised when I saw it move, get up and shake itself. Yes, of course, it was the unhappy Svartflekken who had been in hot water again. He came slowly up to camp, looking very subdued.

But there was an old wolf which had no intention of being frightened off by such a trifle as a few rifle-shots and was not going to let Svartflekken off so cheaply. He started slantwise towards him, whereupon Svartflekken stopped short, faced his enemy, and showed his teeth as well as he could for his muzzle. The wolf was not quite so brave after that. It stopped, and then they both began to walk sideways toward each other, show their teeth and growl, though both were half afraid. Amusing as this might be, I thought we had had enough of it by this time and sent the wolf a bullet through the shoulders, which laid it low. The rest of the pack by this time were in wild flight up towards land, and only two animals were left lying on the field of battle, but of the fugitives Schei had wounded two so severely that they could not keep up with the pack; they turned off into the drift-ice, where they were lost to sight between some hummocks and probably died.

Altogether twelve wolves had attacked the camp. The tracks showed that they had made their assault from two sides at once. My dogs being dispersed, they had made for them first. Schei's dogs, which were lying all together in another place, had probably seemed rather too numerous for a beginning.

Svartflekken was in a terrible state. His head and neck had suffered most, and both his ears were bitten off almost to the roots.

Drip by drip his tracks were red with blood. His face swelled up and at one time he was stone-blind in one eye, and I should say saw very little with the other. On the whole he was not of much use till far on in the season. But just now he was the hero; his fellows sniffed him, licked him, and did whatever they could for him, and he did not object.

On May 20, Sverdrup and Schei set out from the campsite of their wolf encounter and headed for *Fram*, mapping whatever shore they followed as they went. Almost one month later, on June 18, they were back on board ship to find that they were the last of the exploring parties to return. They had added many kilometres of previously unknown Arctic land to the map and had definitely proven that Axel Heiberg was an island separated from Ellesmere Island by Eureka Sound.

Sverdrup and Schei were not alone in adding important new lands to the map of the north. Isachsen and Hassel, who had sledded west, discovered and mapped Amund Ringnes and Ellef Ringnes Islands. It is true that Isachsen and Fosheim had briefly visited the area the previous year, but they had not had the time to do more than ascertain that there was some form of land in view. This time they had mapped the great majority of the coasts of these two bodies of land and had determined that they were in fact two distinct islands separated by Hassel Sound. The Arctic weather station that operated on Ellef Ringnes Island for a number of years after World War II was named Isachsen in honour of the Norwegian explorer.

In addition to the Sverdrup-Schei and the Isachsen-Hassel excursions, two other teams had done extensive mapping closer to home that year for a total of 250 days of traveling and 6,420 kilometres (3,990 miles) covered.

The thoughts of Captain Sverdrup and his crew turned to home, to the warmth of their loved ones and to their native shores from which a full three years of time separated them, not to speak of thousands of kilometres of ocean waters, a good many of which were still frozen solid. In early July, preparations were started to make *Fram* ready for sea. Olsen, the ship's chief engineer, went to work on the engine—tuning, oiling, greasing—for they had to turn many revolutions before the vessel would once more dock in Norway; Raanes, the ship's mate, made new sails and repaired old ones; the forge was kept red hot by Stolz and Nödtvedt fabricating all manner of needed iron fittings and implements; Fosheim applied his carpenter's skills and tools to making and fixing countless wooden accessories on the ship. It would probably be more than a month before *Fram* could break out of her icy prison, but there were many things to be done first. The magnet that was home was pulling very hard and everyone worked with a will.

Not everyone was occupied with getting the ship ready. Although the expedition was nearing its end, exploration and discovery were still uppermost in Sverdrup's mind. Even though his ship was still locked solidly in the ice of Goose Fiord, Jones Sound was open. He set off in a small boat with Schei and Stolz on July 18, across the frigid sound to map the unknown parts of Devon Island on the other side. Here, Sverdrup and his crew found evidence of occupation by Inuit at some time in the past, as he had also found in many places on Ellesmere Island during his three years of exploration.

Although Ellesmere was regularly visited by Inuit from Greenland on hunting trips, the high Arctic islands had not been inhabited on a permanent basis in the recent past. Sverdrup was amazed when he came upon obvious remains of human habitation during his first year of exploration:

> *What was that? A low, ring-shaped stone wall! And still another, and so on all over the point! Here and there were ruined heaps of stones, like beehives in shape, supported by weather-stained, grayish-white bones—the jaws or ribs of the whale.*
>
> *We had, without a doubt, come upon a dead Eskimo settlement. The ring-shaped walls of stone were the remains of their dwellings, and the grey stone beehives their larders. We examined the whole of this dead settlement with great interest. How long could it be since it was deserted? This unexpected arrival amid the marks of human habitation gave us a sudden chilling realization of the loneliness and barrenness of the country. We peeped into the larders; the grass was growing green in them between the stones. We walked from one tent-ring to another; in the centre of each ring was a big tuft of grass, the mark left where the lamp had stood.*

Once again, this time on Devon Island, Sverdrup found signs of habitation—signs that made him wonder if they were not indications of contact between the former Inuit inhabitants and early Norwegians who would have taught them to build shelters to attract eider ducks so they could collect the highly insulating down:

> *The sites of several tents told us, too, that at some time or other, the Eskimos must have been here. As far as I could make out, they had even built nests for the ducks of the same construction that is in vogue to this day in Nordland. At all events we came across a number of very small stone houses. I have never heard that the Eskimos were in the habit of protecting the birds in this fashion yet everything indicated that we were the first civilized people to visit the spot.*

On August 7, Sverdrup and his two companions were back at Goose Fiord. They were not encouraged by what they saw. *Fram* was floating free of the ice, it was true, gently tugging at her moorings, but there was precious little open water anywhere else. The men measured the thickness of the ice down the fiord a way and found more than a metre (a yard).

On August 12, the boiler was fired up and the engine turned over to make sure that all would be in proper working order for when the time came to leave. Sverdrup tried forcing his ship through the ice just to see how much progress he could make through the frozen barrier blocking the way to the open water of Jones Sound. Not very much, he found out. Two or three hours of hard labour grudgingly moved *Fram* forward one or two ship's lengths. Then clear, cold weather set in with night temperatures far below freezing, surrounding the ship with ice so strong by morning that the dogs could walk on it, and in places the men.

On August 16, the weather changed drastically; the frigid north wind abated and was replaced by a warm easterly breeze bringing a rise in temperature to 6°C (43°F). The next day the wind shifted to the south and lanes of open water appeared in the fiord. Sverdrup found that he could bore his ship through the softening ice even though it was still up to a metre (a yard) thick. Then he encountered thicker ice and resorted to blasting in an attempt to open up a channel to the south, but the ice was too thick and blasting was futile.

Sverdrup tried closer to shore where incoming streams had weakened the ice so that *Fram* could steam right through. On August 22, the vessel bored its way through three kilometres (1.8 miles) and if that could be kept up for a few days, the trick would be turned, for it was not that far to the open water of Jones Sound.

The arrival of the neap tides put an end to the progress. The lead of open water next to land was over a shallow bottom that allowed large, broken-up hummocks of ice to anchor there, blocking the ship's way. Until the tides were again high enough to carry the hummocks away, *Fram* was stuck to the spot.

While they were waiting, the crew sawed and blasted a channel south of the ship. Only a few kilometres of barrier remained before the expedition would once more be in open water, steaming and sailing for home.

But it was not to be. On September 5, a cold north wind closed up a promising lead of open water to the south and the ship was stuck solidly in the thickening ice. And there she remained. The boiler was drained and the engine laid up for yet another winter.

The disappointment of the crewmen is not hard to imagine. It was a heart-wrenching blow. They had been away from home for over three years and they were expected back that autumn. What would their wives, children and other loved ones think when they did not return? "Lost like the Franklin Expedition," would certainly enter their minds:

> *The fourth polar night in succession is not a thing to joke about. The worst of all was that we were expected home this year, and perhaps an expedition would be sent out in the summer to look for us.*
>
> *There was another aspect of the matter. How were we to know that we would get free next year?*
>
> *But it was too soon to begin that sort of speculation. If we were to stay here another year we must keep the dogs alive and well through the winter, so that we could do more work in the spring. We had more than enough provisions for ourselves, but had to procure walrus-meat for the dogs at any price, and the sooner the better. However, before we could be walrus-catchers again we had to repair our clothes and footgear, so we set to work speedily with needle and thread.*

Self-pity and desperation were certainly not among the expedition members' weaknesses. If they were to spend another year in the Arctic, it would be a year put to good use.

SOURCE: OTTO SVERDRUP'S *NEW LAND*
The Fram, *frozen in the ice, winter of 1901–02.*

– 6 –

THE FOURTH WINTER

The task of securing food for man and beast was carried out with the usual dispatch and success. Meat was no problem. Sverdrup's greatest food worries centered on what at first glance may appear to be items of luxury, but under harsh Arctic conditions, they assume a more vital character. It is well known that a good cook and satisfying meals go a long way toward ensuring harmony among men working in the conditions that *Fram*'s crew were subjected to: "As for our provisions, what I feared most was that we might run short of butter and coffee. Our after-dinner coffee was struck off the menu, and the butter scales were put into use."

In spite of the harsh conditions and the preparations that had to be made for the coming season of exploration, there was still time for cultural activities:

> *Two aspects of our life during this period remain to be mentioned, one industrial and the other literary. A whole new industry sprang up this winter, for we began to carve in bone. Knife-handles and sheaths of gracefully carved walrus-bone were produced in great numbers, intended as gifts for friends.*
>
> *This was also the golden age of letters on the* Fram. *Bay published a novel called* Gunhild. *Its theme was exceedingly romantic. A party of discoverers went on an expedition to north Greenland; after a difficult journey through the ice-desert they reached a large and fertile oasis where they met descendants of the old Norsemen. These had fallen into two inimical groups,*

continually at war with each other. It was very exciting and was read avidly by all of us.

One thing was uppermost in everyone's mind—getting out of the ice the following summer. Every measure had to be taken to improve the odds of breaking out of Goose Fiord. It is a well-known fact that ice and snow melt much faster under the influence of the thawing sun if they are covered with a dark substance. A familiar example is the case of individual fallen tree leaves boring holes down through the spring snow by the absorption of heat due to their relatively dark colour compared to the surrounding whiteness. Late in the winter, the men spread sand on the ice for a distance of 7–8 kilometres (about 5 miles) to the south of the ship in an attempt to magnify the effect of the coming spring sun's warming rays. Boxes of wood and picks were made up for the job. The sanding went quickly because everyone was anxious to get home and the men all believed in the power of sand to help them.

Plans were drawn up for the spring sled journeys of exploration. Sverdrup and Schei (accompanied part way by Baumann and Raanes as a supply support team) would return through Eureka Sound to Greely Fiord and beyond as far north as they could. Isachsen, Fosheim and Hassel would drive east along the Ellesmere coast of Jones Sound, leaving records of the expedition and building a number of cairns along the way in case a ship should be looking for them during the summer. Seven cairns were built and inside each, in a sealed cylinder, was deposited a message (see facing page), together with a map showing *Fram*'s location.*

When Isachsen came back from depositing the seven messages in stone cairns, he, together with Bay, was to cross over to Devon Island to further survey and map its coastline. When Baumann and Raanes came back from their supply support duties to Sverdrup and Schei, they and Fosheim (also back from his cairn building) would undertake a special sled journey to Beechey Island on the southwestern coast of Devon Island with two specific goals in mind (strange as it may seem since the place was uninhabited): to find out what time it was and to find themselves a sailboat.

On April 1, 1902, the expeditions set off. Sverdrup and Schei, together

---

* The contents of the cylinders are known, as at least five of the seven have been found over the years. Archive searches indicate that they were found by RCMP patrols, the first in 1924 and the last in 1938. Four of the five documents found are in the National Archives of Canada. Two may still be in their Arctic cairns. The message contained in the cylinders, and reproduced on the facing page, has been translated from Norwegian into English.

The Message in the Cairns

NOTICE

From the 2nd Norwegian Polar Expedition with Fram *for any search party that may come into Jones Sound.*

As I cannot overlook the possibility that a ship may be dispatched to look for Fram this coming summer, I am depositing the attached map-sketch of Jones Sound upon which our winter quarters are shown.

As the sketch indicates, Fram *is now in Goose Fiord (Gåsefjorden) where we arrived on September 17, 1900. On account of an exceptionally severe winter and an unfavourable summer, the ice did not break up in the fiord in 1901, when it was my intention to return home. Despite all efforts, we did not succeed in getting out of the fiord, and had to give up trying by September 6. We had moved out 16 kilometres (10 miles) from the bottom of the fiord and had then approximately 8 kilometres (5 miles) to go to open water.*

Fram *can be reached from the edge of the solid ice either by going overland on the east side of the fiord or by going across the ice.*

This notice together with the map-sketch will be placed at the following locations:

1. *Cone Island*
2. *Southern point of the island at the mouth of Harbour Fiord (Havnefjorden)*
3. *South cape (Sydcap)*
4. *Boat Fiord Point (Bådsfjordsnuten)*
5. *Storm Cape (Stormkap)*
6. *St. Helena*
7. *Turn Around Cape (Vendomkap)*

The ice conditions have been favourable in Jones Sound during the years we have been here. Any ship could have navigated the Sound from the middle of July to the end of September.

From our experience, most ice is met with east of Harbour Fiord, while the western part of the Sound is practically ice-free.

On the stretch east of Harbour Fiord navigation should be in the middle of the Sound as there is the least ice there.

Our provisions are plentiful and in good condition, and so we all find ourselves very well.

I have high hopes of returning home with Fram *this fall.*

Fram, *Goose Fiord, the 18th of March, 1902.*
Latitude 76° 39' North
Longitude 88° 59' West of Greenwich
(Signed) Otto Sverdrup

SOURCE: OTTO SVERDRUP'S *NEW LAND*
Starting off on a dogsled exploration trip, spring 1901.

with their support team of Baumann and Raanes, sledded up to the northern end of Goose Fiord and over the neck of land into Norwegian Bay. By April 8, they had reached Hare Point at the entrance to Eureka Sound. The two teams split up here, Baumann and Raanes dropping their extra supplies for the continuing team and returning to *Fram* to undertake their trip to Beechey Island.

The trip by Baumann, Fosheim and Rannes to Beechey Island was a precautionary measure rather than a journey of exploration. Sverdrup was still not absolutely certain of being able to force his ship out of the frozen embrace of Goose Fiord in the coming summer. In case he were stuck yet another year, it would be essential to get a message to Greenland to forestall the mounting of an unnecessary rescue operation à la Franklin. They were not yet in any danger, and Sverdrup wanted to avoid the ignominy of having to be rescued. If necessary, some of his crew could sail to Greenland in one of their small boats to bring the news, but it would be far better to have access to a larger, more seaworthy vessel to get there. It might just be possible to find such a vessel at Beechey Island on the southwestern tip of Devon Island.

It was known that Sir John Franklin had spent the winter of 1845–46 at Beechey Island (which is not an island at all but a peninsula off Devon

Island). When the drive to rescue Franklin began, Beechey became a main location for the searchers to rendezvous and for caching food and supplies in case Franklin came back that way. In 1851, John Ross had left an 11-tonne vessel, the *Mary*, at Beechey for Franklin's use in case he needed it to make his way back to civilization. As late as 1875, an Arctic expedition under Allen Young had visited Beechey and had found the *Mary* still in a seaworthy condition. Sverdrup considered it worth a visit to see if she had weathered well the last 25 years on the beach—and if she had, he could sail her to Greenland.

There was another reason for visiting Beechey. In Sverdrup's day, latitude was easily determined. It was merely a question of observing the highest angle of the sun at high noon, using a sextant or a surveyor's theodolite. It wasn't even necessary to know exactly when noon occurred. It was only necessary to take a number of closely spaced readings around noon to make sure that the sun's highest elevation angle was determined.

Accurately determining longitude, however, was another matter. For longitude, it was necessary to know the time—accurately. Three navigational factors are intimately tied together: longitude, position of celestial bodies and time—the exact time. Know any two and the third can be calculated.

Celestial tables for navigation have been published for hundreds of years. In Sverdrup's day, however, the only way to know the exact time was to carry it with you on a number of highly accurate clocks, or chronometers, from a point where it was accurately known, such as an observatory in a major European city. But even chronometers, accurate as they might have been at the turn of the century, were still only mechanical devices that eventually gained or lost a significant number of minutes over an extended period of time, such as the three years that *Fram* had been absent from Norway at this point. On the south coast of Ellesmere, a chronometer off by 15 minutes meant an error of 64 kilometres (40 miles) in locating a landmark. This was entirely unacceptable for mapping. Sverdrup wanted to adjust his chronometers to the correct time, and this he could do at Beechey Island.

One way of correcting a chronometer 100 years ago was to be at a point where the longitude was accurately known, sighting a known celestial body through a sextant or surveyor's theodolite, correlating the longitude with the angle of the celestial body in navigational tables, and reading the corresponding time from the tables. Because of Beechey Island's strategic location in the search for Franklin, its longitude had been accurately determined by a number of English expeditions with chronometers freshly arrived from Greenwich. By going to Beechey where longitude was known and observing an appropriate celestial body, Baumann and his two companions did indeed correct the three-year deviations of the expedition's chronometers.

The *Mary*, however, turned out to be unserviceable:

> The Mary *was a wreck, although her timbers were well preserved. The deck beams had been sawn off and the deck broken up. The mast had been sawn off about three feet above the deck. In build she was planked diagonally and outside the planking was a sheath of zinc, but on the port side both the sheathing and the planking were very badly damaged. One lifeboat was still there, but it was not seaworthy.*
>
> *The general impression was one of wanton destruction, but whether Eskimos or whalers had been the perpetrators, it was impossible to tell.*

So much for the *Mary*. If Goose Fiord kept *Fram* prisoner again the next year, Sverdrup would have to rely on one of his own small boats to sail to Greenland and let the world know that he and his crew were still alive. No one really wanted to envision another winter in the ice, but it was better to plan ahead, just in case.

The team of Sverdrup and Schei continued up Eureka Sound, encountering little game, but finally killing a polar bear on April 16 to feed both men and dogs. It was to be the last big game they would kill until May 17, while on their way back.

By April 28, they camped a few kilometres north of Butter Porridge Point, the most northerly point they had reached the previous year. From that time on, they were exploring and mapping the unknown western coast of Ellesmere Island and reaching for the most northerly point that the expedition would attain during its four years of Arctic exploration. On May 7, Sverdrup and Schei stood at 81° 41' north latitude and named the place Lands Lokk (meaning Land's End), not only the furthest north point of the expedition—some 920 kilometres (572 miles) from the North Pole—but also the most westerly point of Ellesmere Island. Sverdrup and Schei made some butter porridge at Lands End, a dish, by then become traditional, which they enjoyed at the northernmost point of each sled journey.

On May 9, Sverdrup and his companion started for home, cutting across the sea ice to the northern tip of Axel Heiberg Island and then down the eastern coast of the big island, reaching Butter Porridge Point on May 16. Now they were in serious need of food, especially for the dogs. They had shot the last big game exactly one month before and that meat was long gone. The dogs had been living on dried fish and biscuit for too long. They were getting weak.

It was most appropriate that Constitution Day, May 17, should see a herd of muskoxen fall to the bullets of Sverdrup and Schei in their time of great need. The celebration could now be done properly, given the circumstances, although without the usual dram or two of spirits which had been exhausted by past celebrations:

> *In the afternoon we spread a large ox-skin on the ground and pitched the tent on it, and placed another large skin outside the door so that we could take off our boots and be comfortable in the open air. There was brilliant sunshine and only a few degrees of frost. While the cooking was going on the tent-door stood wide open, and after a sumptuous meal of broth, meat, marrowbones, and other delicacies we lighted our pipes and lay comfortably in front of the tent, half-hidden in the long, soft hair of the skin, while our pretty Norwegian flag waved from the roof.*
>
> *What a change! Weeks of toil and hardship, and anxiety over the fate of our dogs, and now here we were steeped in unadulterated well-being. We had not a single drop of spirits of any kind, but we really did not need any. A pleasanter Seventeenth of May would be hard to find, and the dogs shared the joys of existence with us; they reveled in the delicious warm meat until they could hardly move their jaws.*

Down Eureka Sound sledded Sverdrup and Schei, day after toilsome day. Upon reaching Bay Fiord, they decided to explore it to its end, and there a great temptation seized them.

During the expedition's first winter in Fram Haven, Sverdrup and Bay had first discovered this body of water by sledding up Flagler Fiord and then crossing overland via Sverdrup Pass. It was a relatively short distance to reverse the order and find themselves back at Fram Haven with Pim Island just across the narrowness of Rice Strait. The reason for their temptation lay on Pim Island, for it was there that Peary's supply ship had left mail from Norway for them in August of 1899—mail that insuperable ice conditions had prevented Sverdrup and his men from picking up at the time. True, it was mail that was almost three years old, but it was still mail from home, and that was precious beyond measure. But reason got the best of emotion. The season was far advanced and Sverdrup would be lucky not to run out of snow and ice even if he took the shortest route home, so he and Schei turned their backs on their dreams and continued south down Eureka Sound.

The spring ice began to badly cut the dog's feet. Sverdrup made canvas booties for them, but they were only partially successful. By June 15, the party reached the neck of land separating Norwegian Bay from Goose Fiord and by early afternoon the next day, were back on board *Fram* after an absence of 77 days. The four main sled expeditions during the fourth year of exploration had spent 168 days on the trail and had traveled 4,240 kilometres (2,635 miles).

All was well on board ship, and what especially pleased the captain was the broad-sanded street leading down the ice almost eight kilometres (five miles) south of the ship, ready to help the sun in its spring work of destroying the ice. The sand did its job well. On July 15, *Fram* finally floated free of the imprisoning ice and could have sailed down in clear water to the end of where the sandy belt had been, for a belt of open water had replaced it. But the 24 kilometres (15 miles) of fiord remaining from that point on before reaching Jones Sound was still solidly locked in ice.

The next three weeks were an emotional roller coaster for Sverdrup and his crew. The previous year's failure to break out of Goose Fiord was uppermost in their minds:

> *This Goose Fjord is a very remarkable place. It had to wait a long time before strangers found their way into it, but once they were there it knew to perfection the art of keeping its guests. Last year the ice never broke up at all; this year, when the same host opened the drawing-room for his guests to leave, he seized them by the collar and held them fast in the hall. "I will never do this again," said the boy as he chopped off his left hand. I, at any rate, will take good care not to set foot in Goose Fjord again.*

Another serious matter preoccupied Sverdrup. On July 7, Bay, Simmons, Isachsen and Hendriksen had set out in a small boat to do some dredging for sea animals and plants in the direction of Cardigan Strait. By July 30, they were already nine days overdue. This was truly worrisome. If *Fram* did manage to extricate herself from her trap, there was no question of leaving without them. They would either have to be found, or their fate determined. Time spent searching, though, could very well make the difference between sailing home and spending a fifth year in the ice. But the Arctic was not yet ready to release her guests and she started getting a bit nasty about it. Sverdrup and his men had to keep a sharp watch to make sure they did not lose their ship to the grinding blocks of ice surrounding them:

> *Just before midnight on 2 August Hassel, whose watch it was, came running down and turned me out with the news that an enormous floe was bearing down on us. The rest of the mate's watch were immediately turned out and the second anchor got ready in case the chain we had out broke.*
>
> *The floe came sailing along, gave us a slap, and swept us inwards with irresistible force. It hooked itself into us securely, and seemed to have no intention of letting its catch go. There were not many feet of water under the keel when its grip slackened and we were freed. Between three and four in the morning, when the water was falling, the same floe came slinking down the fjord again, caught us once more, and out we went. This time it was bearing straight down on the peninsula just abreast of us, and nearer and nearer it came.*
>
> *Well, the* Fram *would probably stand this shock too, but just the same we had to be at our posts. Both watches on deck! Hawser round the big ice-blocks stranded over there! When the hawser was taut we paid out some more chain and got the ship clear of the floe and closer to land.*
>
> *When the floe, having failed in its mission, had floated past, the engineers were turned out and with all hands on deck we began to heave off. We were tired of being in such an exposed position; nor would it do to be here at the time of the spring floods, so it was best to move on.*
>
> *We moored the ship by the stern, close inshore and with only a couple of feet under the keel at low tide; we would now have the bottom as a refuge in case of dire necessity. We put a stout steel cable round a large rock up on land, and here we could at least lie in peace without risking anchors and chains.*

On August 5, a cry went up. Schei had spotted movement on the distant shore through his binoculars. The long-awaited four dredgers had returned! A shiver of joy and relief spread through the crew at the safe return of their companions. For 10 whole days they had been stranded by the unyielding ice on a small island in Cardigan Bay. The buffeting wind and rain and the prospect of starvation on their outcrop of rock incited them to name it Devil Island. But now they were back and *Fram* could leave at any time if only the ice would let them.

The return of the dredgers could not have been more timely. The same afternoon the north wind began blowing in earnest, clearing the fiord of ice. By the next morning the boiler was fired up and the moorings taken on board. At 11:00 a.m. the little vessel steamed out of Goose Fiord into Jones Sound, free at last and under full sail, homeward bound!

- 7 -

HOME TO NORWAY!

FREE OF THE ICE, YES, BUT THE ARCTIC had other tricks up its sleeve. A squall battered *Fram* as she shouldered her way through the rough seas and split her jib. Luckily, she was just outside Harbour Fiord, her second winter quarters, so she quickly ran in to the shelter of her former anchorage to make repairs. The next day, when all the work was done, Sverdrup went ashore to visit Braskerud's cross, which had been raised above the spot where his body was laid to rest in the cold Arctic waters. An inquisitive bear had pawed away a few stones at the base and the cross was leaning to the side a bit. Sverdrup straightened it as best he could, paid his last respects to his old friend and left the place forever.

On August 9, Sverdrup and his crew headed out into Jones Sound; destination—coast of Greenland, through waters essentially unencumbered by ice except for the occasional, but very dangerous, iceberg. Eight days later *Fram* anchored at Godhavn, Greenland, back in relative civilization for the first time in over four years. The crew looked forward to fresh mail from home; all but Simmons and Bay were destined to be disappointed. It was certainly not indifference on the part of their loved ones back in Norway that accounted for the paucity of mail. There was lots of mail, but as luck would have it, it had been sent further up the coast of Greenland to Upernavik.

The crew's memories of Godhavn were both happy and sad: happy because the people of the small Greenland community treated them like royal guests during their three-day stay, making their reacquaintance with civilization as pleasant as possible; sad because they had to leave behind their hard-working and faithful traveling companions—their dogs. Those that were still young and in good shape were no problem; Sverdrup gave them away mostly to the settlement superintendent and the pastor. The older ones which did not have much work left in their tired bodies were another matter. Dogs in the Arctic had to pay for their keep in the only way they knew how—by working. Sverdrup knew only too well what happened to dogs that were too weak or too old to work—they did not get fed and they starved to death in misery and pain. He could not face this possibility, so he had them shot. As contradictory as this may seem, his decision was made out of consideration for their ultimate well-being. The poor creatures were quickly disposed of:

> *The animals were as fat as they could be and there was eager competition among the Eskimos for the remains; they skinned them and feasted on the bodies.*

Sverdrup brought a few of his Greenland dogs back to Norway with him, where they became the foundation for a new breed of dog.*

On August 21, *Fram* departed Godhavn amid saluting salvos of gunfire from shore and answering bursts from the ship's own two cannons. Ten days later the vessel was about 320 kilometres (200 miles) southwest of Greenland's southern tip, Cape Farewell, when chief engineer Olsen brought his captain disquieting news—the engine had suffered serious damage. He could only count on it for service in a pinch for short periods of time from then on, and even at that, only with low pressure in the steam boiler. This was a frustrating development for everyone; the captain and his crew were all desperately anxious to be reunited with their loved ones again. But it was not a crippling development. The winds were favourable at that time of the year and the vessel had a full complement of sails. In fact, Sverdrup was quite thankful that the problem had not surfaced earlier when he had really needed steam power, while bulling his way through the imprisoning ice of Goose Fiord.

On September 18, Sverdrup sighted the island of Utsira off the southwest coast of Norway and the next day obtained a pilot from there to guide *Fram* into the nearby port of Stavanger—and here the celebrations began.

The town authorities of Stavanger boarded *Fram*, all except His Worship

---

* During World War II, all Greenland dogs in Norway were requisitioned by the occupying Germans. After the war, new dogs were imported by Norway from Greenland to re-establish the breed.

the Mayor, who did not trust his rope-ladder climbing abilities to see him safely aboard. The townspeople of Stavanger crowded the dock shouting, jumping and excitedly waving welcome to the Arctic travelers, who had almost been given up for lost.

The distance from Stavanger to Oslo is about 640 kilometres (400 miles), but it took *Fram* 10 days to sail it, not because of her broken engine or lack of wind, but because of the celebrations and ceremonies along the way. Sverdrup had to decline many invitations from towns along the way to stop. Had he accepted them all, there is no telling when he would have entered his home port of Oslo:

> *We had been met by quite a fleet of steamers and sailing-boats as far out as Horten, and the* Fram's *triumphal procession from Stavanger to Oslo ended on a beautiful autumn Sunday which reminded us of the day four years before when we had gone the other way. What a difference between then and now, and yet how near each other these days seemed to us! It was as if the frost and ice of the polar night melted away before all the warmth that flowed out to greet us in the welcome of our countrymen; as if the remembrance of the four long years with all their toil was buried under the sweet-smelling flowers that were showered over us as we drove through the streets of Oslo; as if all the waving flags could spirit away the furrows the winter had brought us.*

Sverdrup's achievement is unparalleled in the history of Arctic exploration. During the four years of his expedition, 15 major sled journeys were undertaken, making a total time on the trail of 762 days and covering 17,515 kilometres (10,885 miles). There were also a great number of shorter trips of 5 to 20 days covering from 80 to 480 kilometres (50 to 300 miles) each. In all, he and his men explored some 260,000 square kilometres (100,400 square miles) of previously undiscovered land, mapped it, and took possession of it in the name of his country.

In addition to discovering, exploring and mapping vast tracts of previously unknown Arctic lands, the second Norwegian Arctic expedition in the *Fram* produced a wealth of other scientific data and artifacts. Several of the members of the expedition were scientists and made important collections of specimens in their respective fields: Simmons was a botanist, Bay was a zoologist, Schei was a mineralogist, geologist and paleontologist. Baumann had studied electricity and performed magnetic observations. A series of meteorological observations was also compiled.

Amongst the more interesting discoveries were the very well-preserved, 45-million-year-old remains of species of sequoia and cypress trees that today only grow in China. The climate of the Arctic eons ago was evidently vastly different from what it is today.

The wealth of scientific data and collections amassed by the expedition gave rise to intense scientific study in the years following Sverdrup's return. The results of these studies were published in five volumes: two in 1907, one in 1911, one in 1919 and a supplementary volume in 1930. In all, the five volumes comprise over 2,000 pages of material.

PART II

THE CANADIAN EXPEDITIONS 1903–1948

– 8 –

CANADA IS NOT AMUSED

OSLO AND ALL OF NORWAY WERE ECSTATIC at Otto Sverdrup's amazing discovery and exploration of new land in the High Arctic. Celebrations erupted in the streets. On the other side of the Atlantic, however, the news of Sverdrup's discoveries and claims in the name of Norway hit Ottawa like a bomb shell. The Canadian government was aghast, caught completely by surprise.

From the time of Martin Frobisher's first venture into the Arctic in 1576 until the mid-19th century, the discoverers of the High Arctic had been, with few exceptions, British. By the second half of the 19th century, some Americans had also left their marks on the northern map, but by far the great majority of the names were from the British Isles—Frobisher, Parry, Jones, Lancaster, Ellesmere, Franklin, McLure, McClintock, Belcher, Ross.

All the discovered land north of mainland Canada that was not part of Greenland had been claimed by the British Crown. In 1880, by Order-in-Council, England had given these Arctic lands to the young Dominion of Canada to watch over and protect for the British Empire, especially against American intruders. However, there was no apparent danger of challenge to the ownership of these lands at that time, so Canada fell into a period of complacency. Few efforts were made by Canadians to even visit, let alone explore their far northern domain. The age-old principle of discovery settled ownership in those days, and the British had indisputably been the discoverers. But now, at a stroke, a rash of new, foreign names intruded on the formerly empty space north of Canada—Sverdrup, Axel Heiberg, Ellef Ringnes, Amund Ringnes, Björneborg, Svendsen, Steinkjer. Foreign land between Canada and the North Pole? The Laurier government in Ottawa was stunned.

Sverdrup's new discoveries and claims in the name of his country shocked Canada out of its complacency. In fact, there were other threats to Canadian sovereignty in the Arctic besides the Norwegians, but it took Sverdrup to stir Canada into action. American whalers cruised the frigid northern waters at will, asking leave of no one and paying no fees for the privilege. Inuit from Danish Greenland regularly crossed the narrow gap of Smith Sound with impunity to hunt musk oxen on Ellesmere Island. The very year after Sverdrup's triumphant return to Norway to a hero's reception, the belated era of Canadian Arctic patrol and exploration began. Sovereignty in the Arctic had suddenly become a serious preoccupation of the Laurier government.

Canada lost no time. In 1903, two groups of Canadians pushed north to defend their country's sovereignty, one at each end of the country. In the west, the Royal North West Mounted Police (RNWMP, later to become the RCMP) created two Arctic detachments—one at Herschel Island in the Beaufort Sea near the Alaska–Yukon border, and one at Fort McPherson in the North-West Territories south of present-day Inuvik. The officers sailed to their postings on board ships chartered from private companies. The RNWMP did not yet have its own ship, but this would change in time. The officers' duties were to demonstrate a Canadian presence in the Arctic by, among other things, collecting license fees from American whalers for the privilege of harpooning the huge marine mammals in Canadian waters and collecting duties on goods taken out. Until that time, American whalers had made themselves completely at home in Canada's north. At the other end of the country, the annual Eastern Arctic Patrol first sailed to the frigid waters of the north in 1903, also to show the Canadian flag and to control the eastern whaling fleet.

– 9 –

A.P. LOW AND THE NEPTUNE, 1903–1904

ALBERT PETER LOW (1861–1942) was one of Canada's foremost geologists at a time when practicing that profession required the hardened physique of a *coureur de bois*. He was ranging northern Quebec and Labrador for the Canadian Geological Survey (GSC) at almost the same time Sverdrup and his crew were discovering new land in Canada's Arctic archipelago at the turn of the century. Low put a vast area of wilderness Canada on the map in the late 1800s. His explorations and surveying in the years from 1893 to 1895 laid the groundwork for the definition of the rugged Quebec–Labrador border.

Low was no ordinary geological surveyor. He was a man of great physical strength, endurance and conviction. Reading about Low in F.J. Alcock's 1947 book, *A Century in the History of the Geological Survey of Canada*, helps one get a measure of the man:

> *In 1884 a joint federal and provincial expedition was sent to Lake Mistassini at the head of the Rupert River, with John Bignell, a Quebec surveyor, in charge, and with Low responsible for the geological investigations.... The greater part of the distance was covered by canoe and it was only after a long and difficult tramp on snow-shoes, during the last 10 days of which rations were very low and the temperature ranged to 40 degrees below zero, that the Hudson's Bay Company post on Lake Mistassini was finally reached on December 23.*

> *The party was to spend the winter in the region and continue work the following year. Certain disagreements arose, however, between Low and Bignell regarding the operations of the party, and to clear up the matter Low packed his toboggan, put on his snow-shoes, and started for Ottawa, which he reached on March 2.* On March 23 he received instructions to return to Mistassini in full charge of the party. He left the following day, and after another difficult journey arrived there on April 29. He completed the survey of the lake and in the autumn descended the Rupert to James Bay.*

A round trip of 87 days mainly of backbreaking labour, just to untangle an administrative problem. In a talk given by Low in 1890 on his work surveying Lake Mistassini, Low laconically limited his comments on the incident to "Bignell was recalled in the spring of 1885," not even mentioning the trip to Ottawa. However, Low did describe the trip in some detail in the report he wrote on it for the GSC. This report gives some insight into the appalling conditions under which GSC geologists had to work in those days:

> *Two heavy snowstorms occurred while we were on our way, making walking so difficult that our tent and sheet iron stove had to be abandoned, and we were obliged to sleep in the snow for more than a week.*

In 1903, the year after Sverdrup's triumphant return to Norway, Low was given command of the 420-tonne Newfoundland sealer *Neptune* with orders to

> *patrol the waters of Hudson Bay and those adjacent to the eastern Arctic Islands; also to aid in the establishment, on the adjoining shore, of permanent stations for the collection of customs, the administration of justice and the enforcement of law as in other parts of the Dominion.*[1]

Low's vessel was the largest and most powerful of the Newfoundland sealing fleet. She was a three-masted schooner carrying a coal-burning steam engine as well as sails. Her 550-horsepower engines could thrust her through the northern sea at a steady eight knots on a calm surface. *Neptune* was already solidly built, but she was even further reinforced to stand up to

---

* In fact, Low snowshoed from Mistassini to Lake St. Jean, continued on by horse and sleigh to Quebec where he boarded a train for Ottawa.

SOURCE: NATIONAL ARCHIVES OF CANADA (C-1764)
The Neptune, *all dressed up for July 1, 1904.*

the crushing ice pressures that go with Arctic sailing. Her sides were sheathed with nine centimetres (four inches) of greenheart, just as *Fram's* had been. Under that came a four-inch layer of oak planking on top of a frame of heavy oak timbers. The inner sheathing of the oak frame was another 7.5 centimetres (three inches) of oak planking. The hollow space between the inner and outer hulls was filled with rock salt to absorb and distribute the deadly, crushing pressure of the inevitable ice floes she would encounter.

On August 23, 1903, *Neptune*, captained by S.W. Bartlett of the renowned Newfoundland seafaring family, quietly slipped her Halifax berth with Low and a contingent of 35 others on board, including one RNWMP noncommissioned officer and four constables under the orders of Superintendent J.D. Moodie.

Low's orders were to find and pass the winter in the company of an American whaler, the *Era*, known to be somewhere in Hudson Bay. American whaling ships used a number of harbours in the northwest corner of Hudson Bay to lay up for the winter, but Low was not sure which one the *Era* would be in.

SOURCE: NATIONAL ARCHIVES OF CANADA (C-33700)
The Neptune's *crew in winter dress.*

It was clear that the question of who owned the north was far from settled. Moodie and his men were under orders to build the first of planned, permanent police posts at a suitable location, from which they could control whaling activities in Hudson Bay in the name of the Canadian government. There was no longer to be any doubt about Canada's claim of sovereignty, at least in Hudson Bay.

Adding spice to the sovereignty game, just a little over a month before Low sailed north from Halifax yet another Norwegian, Roald Amundsen, had sailed from his country to Canada's eastern Arctic on board the 47-tonne sloop *Gjøa*, bound for the northwest passage. He wintered over twice at Gjøa Haven on King William Island before finally emerging on the west coast of North America in 1906, becoming the first captain to con his ship through the famed passage.

Neptune sailed up the full length of the coast of Labrador to Baffin Island, then came back south and in to Hudson Strait and continued all the way west to the vicinity of Chesterfield Inlet, stopping at various outposts along the way. While sitting out a gale at Depot Island near Chesterfield, *Neptune* was visited by several Inuit who told Low that the *Era* was

SOURCE: G.F. CALDWELL, NATIONAL ARCHIVES OF CANADA (PA-29001)
A.P. Low and officers, 1903–1904. Low is in front row, centre.

anchored some miles farther north along the coast at Fullerton Harbour. Before heading for Fullerton to spend the winter, Low decided to take *Neptune*'s launch with a crew of five men and an Inuk pilot into Chesterfield Inlet up towards Baker Lake, where his Inuit visitors were hunting barren-ground caribou. There he would trade tobacco, knives and files for meat and caribou skins for making winter clothing. Meanwhile, he sent Captain Bartlett with the rest of his crew and another Inuk pilot to guide the *Neptune* into Fullerton Harbour where the *Era* was wintering over.

Low and the *Neptune*'s launch ran into serious trouble in Chesterfield Inlet. They were successfully guided up the Inlet and completed their trading as planned, but on the way back down with some hundred skins and a considerable quantity of meat, the Inuk pilot became confused in the gathering dusk and recommended anchoring for the night off aptly named Dangerous Point. An anchor watch was set and the rest of the crew went to sleep. At midnight, the alarm was raised by the watch. The launch was aground and the tide was falling. Efforts to refloat her failed and she keeled over on her side with water pouring in. Some of the crew was set to saving the cargo by transferring it to a nearby island while others bailed frantically to try to refloat the launch, but to no avail. At high tide the following morning efforts to refloat the craft were unsuccessful, so Low decided to send the launch's 4.5 metre (14-foot) dinghy north up the coast with two men and Scotty the Inuk pilot to get emergency help from the *Neptune*, which by then should have reached Fullerton Harbour.

On September 23, the very day Low and the launch left her, *Neptune* had sailed into Fullerton Harbour to be enthusiastically greeted by Captain George Comer of the American whaler *Era*, who was already there, preparing to be frozen in for the winter. Three-and-a-half days later, the dinghy and its dead-tired crew of three reached *Neptune* in Fullerton Harbour and raised the alarm about the sinking of the launch. *Neptune* sailed to the rescue.

On the afternoon of October 3, Low and his shipwrecked sailors were heartened to spy smoke from the stack of the *Neptune* arriving to succour them. The following morning, three small boats from the ship accompanied by Captain Comer's whaleboat arrived at the scene of the wrecked launch. It was not until October 8, with a three-day gale intervening, that *Neptune* itself finally reached the scene of the sinking and hoisted the launch aboard, revealing large holes in both sides. Another gale on October 10 again immobilized operations and it was the following morning before *Neptune*'s anchor was finally raised and arrived safely at Fullerton that evening at dusk, launch and all, 18 days after the sinking.

For the next nine months, *Neptune* and *Era* spent the winter locked in the icy prison of Fullerton Harbour. Low and Comer became the best of friends, with the American captain imparting much of his hard-earned knowledge of the area as a whaling captain to a very grateful Low. Comer willingly paid his whaling fees.

The winter was not quite uneventful as there were two unfortunate deaths among *Neptune*'s men. From Low's book, *The Cruise of the Neptune*:

> Dr. Faribault had shown signs of mild insanity, almost from the time of leaving Halifax. On the 1st of November he became violently insane, when, on the advice of Dr. Borden, he was placed in charge of the police as a dangerous lunatic. The poor man had to be confined to a cell, and watched continuously. His condition became worse and worse, until he was happily released by death on the 27th of April following.[2]

RNWMP Superintendent Moodie made the following entry in his trip report:

> *The accommodation in the guard room was necessarily very limited (this and the cell being the same size, 6 feet by 8 feet [2 metres by 2.5 metres]). With the yells, at most times, of a raving maniac ringing in their ears through the grating of the cell door two or three feet [less than one metre] away, it can easily be understood how trying was the situation. In spite of everything, their good temper and kindness in the handling of the patient never lessened. This description of duty for so long a period scarcely comes within police work, and I would strongly recommend that 25 cents per day additional pay be allowed for it, say under the heading of "Attendant in Asylum."*

One can imagine the effect of this situation on the men on board ship in the lonely cold and darkness of the subarctic winter.

The second sad occurrence that winter was the death on December 11 of cabin boy James O'Connell, described by Low as being of "weak mind." O'Connell left ship to visit Inuit snow houses which had been built on shore to be near the ships. After he left, a sudden blizzard came on as is not rare in Arctic and subarctic regions, but his absence was not noticed until the blizzard developed into a whiteout gale and searching for him proved impossible. The storm raged for two days before some Inuit found his tracks leading to open sea water some five kilometres (three miles) from ship. The only consolation to lighten the pall of sadness as expressed by Low was that, "There is no doubt that death came quickly."

On July 18, 1904, after lying nine months trapped in the solid grip of Hudson Bay ice, Low finally succeeded in winching up his anchors and sailing east for Port Burwell on Killinek Island at the very tip of the Labrador coast. He had a rendezvous at Burwell with the relief ship *Erik* to take on food and supplies from her for his coming northern mission, which would take him high up into the Arctic Islands. At this point, Superintendent Moodie climbed over the gunwale of the *Neptune* and transferred to the *Erik* for the southbound sail to Ottawa.

On August 2, a resupplied *Neptune* boldly sailed north into Smith Sound, which narrowly separates Ellesmere Island from Greenland, to formally take possession of Ellesmere for Canada, whose western half Sverdrup had just discovered, explored and claimed for Norway.

Low and his men set foot on the eastern coast of Ellesmere at Cape Herschel on August 11, 1904, and took formal possession of the whole

SOURCE: NATIONAL ARCHIVES OF CANADA, A.P. LOW (PA-53580)
A.P. Low and crew at Beechy Island cenotaph erected in memory of the perished explorers of the 1852 British Naval Expedition commanded by Sir Edward Belcher.

island for the Dominion in the name of Edward VII. He proudly raised the Canadian flag, read a proclamation of claim and customs regulations and left a copy of it in a large cairn of piled-up stones for all passersby to visit and examine as physical evidence of Canada's claim. Low then turned south into Lancaster Sound, which separates Devon Island from Baffin Island, to lay claim to more Arctic islands with displays of the required pomp and circumstance at Port Leopard on northern Somerset Island and at Beechey Island on the southwest coast of Devon Island.

Beechey Island has special significance for Arctic sailors. It is here that the ill-fated Franklin expedition in 1845–46, during the initial winter of its agony, offered up its first victims to the cause of the search for the Northwest Passage, only to subsequently disappear itself into the icy Arctic mists, never to be heard from again. It is at Beechy that the expedition buried the bodies of three of its crew members who died that first winter. The 140-year-old corpses were eventually exhumed in 1984, almost perfectly preserved by the permafrost.[3]

On Beechey Island, a cenotaph rises above the rocky Arctic soil. It keeps a lonely vigil in memory of the perished explorers of the 1852 British Naval Expedition commanded by Sir Edward Belcher. A second memorial, in the form of a marble tablet later left at the same spot by Captain McClintock on behalf of Lady Franklin, honours the memory of her late husband and his perished crew.

When the *Neptune* first arrived at Beechy Island, Low had found a tin case attached to the cenotaph with a record in it left by Roald Amundsen of the Norwegian Magnetic Pole expedition dated August 1903, almost exactly one year earlier. Amundsen was a famous polar explorer who in 1911 later would become the first to reach the South Pole, beating the ill-fated Robert Falcon Scott team by about a month. In 1904, however, Amundsen was on his way to achieving another first, in the Arctic this time, of being the first man to sail through the Northwest Passage. His ship was the tiny, 43-tonne sloop *Gjøa* and he had stopped at Beechy Island to leave a record of his progress to date in case he ran into trouble, or did not survive. Low removed Amundsen's record and later forwarded it to the Norwegian government. In leaving Beechy Island, Low sailed across Barrow Strait to Port Leopold on Somerset Island, again finding evidence of Amundsen's expedition in the form of a cache of provisions left there for him by the whaler *Windward*.

His work done, Low turned the *Neptune* toward home. On his way, on October 1, he dropped anchor at Port Burwell at the very northern tip of Labrador where two months before he had been resupplied by the ship *Erik*, coincidentally meeting Captain Joseph Bernier on board his ship, appropriately named the *Arctic*, heading north to become the next Canadian to respond to the Norwegian threat to Canadian sovereignty in the north. Superintendent J.D. Moodie was on board with Bernier, heading for the Fullerton Police Post to over-winter once again in the north, this time in the company of his wife and young son. Scarcely more than two months before, Low had dropped Moodie off at Port Burwell. Moodie had then sailed south to Ottawa on board the resupply ship *Erik*. Moodie barely had time to repack his bags, gather his family and jump onto Bernier's northbound *Arctic*, headed back to Fullerton Harbour.

Fifteen months after he had sailed north in 1903, Low dropped his anchor in Halifax harbour on October 7, 1904, after logging 16,000 kilometres (10,000 sea miles), his job well done.

– 10 –

JOSEPH ELZEAR BERNIER, 1904–1925

IN 1904, CAPTAIN JOSEPH ELZEAR BERNIER (1852–1935), a burly, no-nonsense sailor, took up where Low left off. Bernier was born on the shores of the St. Lawrence River at L'Islet, Quebec. His was a family of sailors: father, grandfather and two brothers were sea captains, two other brothers were river pilots. Joseph's father took him to sea from the age of two. In 1869, when he was 17, he was made captain of his father's brigantine, the *Saint Joseph*, and sailed a cargo of lumber from Quebec City to London in her.

The captain of a ship had to be strong in those days, strong of character and strong of physique. At 17, Bernier was both. The exercise of authority and the administration of justice were quick and rough on board ship, but they worked. More than once in his career, Bernier had to physically wrestle malcontent and often drunken seamen to the deck as they tried to challenge the captain's authority. Bernier once sentenced a mutinous and lazy sailor be kept in irons and be fed bread and water once a day. In his memoirs Bernier wrote:

> *On hearing the sentence meted out to him, the man was beside himself with rage. Breaking loose from the men who held him he attacked me savagely. I secured a firm hold on him, pinioning his arms; but he tried to bite my arm and I was forced to strangle him into submission.*
>
> *But this man was truly stubborn. He did not consider himself beaten by this punishment. During the night ... he would begin*

hammering and kicking the door, renewing the noise often enough to keep anyone from sleeping. So I had him taken out, had a hole bored in the deck in the centre of the room and a ring-bolt fitted into it, and the man chained to this ring. Here he was locked up again and given his ration once a day of bread and water. The second day he told the sailor who brought him his ration that he wanted to see the captain. So the mate and I went in with the logbook, and his statement that he was now ready to do his work was duly entered. He was then released with the warning that if he made any further attempts to arouse the men to mutiny that he would be put in irons for the remainder of the trip and handed over to the police on arrival in port. In this way was incipient mutiny checked in those days of "wooden ships and iron men."[4]

Two years later, in 1871, Bernier was docked on the Potomac River near Washington, DC, when he saw Arctic explorer C.F. Hall's ship *Polaris* in drydock being readied for her final trip to northern waters from which she was never to return, crushed by the Arctic ice. Bernier foresaw the disaster as he examined the ship's lines with a critical and knowing eye. She was built too straight-sided, not round-bottomed enough—certain to be crushed by tonnes of irresistible icy pressure, instead of being squeezed upward and out of the deadly vise, like the *Fram* was, by this same pressure. He predicted disaster, but his warning was ignored and the *Polaris* soon sailed to her icy doom.

Bernier developed a passion for the Arctic, a passion that would burn in him for life. The glory of being the first man to discover the North Pole beckoned to Bernier. He was well aware of the cross-polar ice drift that had carried the wreckage of the American ship *Jeannette*, which had been crushed off Siberia, to where it was discovered off the coast of Greenland several years later. He was also fully aware of Nansen and Sverdrup's tantalizingly close approach to the pole by forcing *Fram* into the ice-pack and letting her drift for three years. In fact, Bernier wanted to do almost exactly what Nansen and Sverdrup had tried and failed, only he would drive his ship into the ice further east, north of Bering Strait, hoping to compensate for the flaw in the Norwegians' plans that caused them to miss the pole by just a few hundred kilometres.

Personal challenge was Bernier's driving force, but Canadian sovereignty over its northern domain was also very clearly in his plans:

Why should Canadians not go as far north as ninety degrees and place their flag on that part of the globe, the northernmost boundary of Canada, a country that is part of the British Empire.[5]

For a time, it looked like Bernier might realize his dream. He went to Germany and sailed back with the *Gauss*, a full-rigged sailing vessel and one that had proven her worth in icy waters. She had wintered over in Antarctica in 1902–03. Bernier sailed her back to Quebec City where she was renamed the *Arctic* and immediately fitted out for a northern expedition.

Canadian sovereignty was also very much in the mind of Wilfrid Laurier and his government, but unfortunately for Bernier and his dream, the prime minister thought there were better ways of protecting Canadian sovereignty than by trying to push through to the North Pole. Bernier received sailing orders that had nothing to do with the pole:

... the Arctic *will be under the command of Captain Bernier and ... is to sail on August 15 [1904]. This boat will carry an officer and ten men of the mounted police, apart from the crew of the ship... Their instructions are to patrol the waters, to find suitable locations for posts and to assert the jurisdiction of Canada... At the present time there are whalers and fisherman of different nations cruising in these waters, and unless we take active steps to assert ... that these lands belong to Canada, we may find ourselves later on in the face of serious complications."*[6]

Bernier may finally have been responding to the siren-call of the north, but it was not under the conditions that he would have wished. His dream of drifting across the North Pole was not to be realized, and to top it all off, Bernier was merely called upon to act as a ship's captain on this trip—RNWMP Superintendent Moodie was to command the expedition. One can judge the measure of the man by his noble reaction to a situation that could only have been deeply disappointing:

... since the opportunity to make a drift across the North Pole was to be denied me, I was determined to devote my efforts in the Arctics [sic] *to what after all may be regarded as a more important object, that is to say to securing all the islands in the Arctic archipelago for Canada.*[7]

In spite of Laurier's order to sail on August 15, it was mid-September 1904 before Bernier and *Arctic* stood out of Quebec City harbour and sailed

SOURCE: NATIONAL ARCHIVES OF CANADA (PA-20904)
The Arctic, *moored at Quebec City in 1904.*

north, meeting up at Port Burwell with Low, in the *Neptune*, who was on his way home. On October 16, *Arctic*, with Bernier at the helm, sailed into Fullerton Harbour and was welcomed, like Low in *Neptune* in 1903, by Captain Comer of the American *Era*, wintering over for his third year in a row.

Comer was sociable, and his relations with Bernier were as cordial as they had been the year before with Low. He had whaled in northern waters for many years now, and had valuable information to share with Bernier about sailing that part of the globe. Bernier, Comer and the Moodies enjoyed many meals together. Socializing was essential to while away the long hours of northern darkness. The captains hosted dances on board their ice-imprisoned ships and the guests came decked out in both southern and northern fashions. Tattooed Inuit women proudly displayed their exquisite bead-embroidered finery as they swung their partners under the cold and dark northern sky.

SOURCE: NATIONAL ARCHIVES OF CANADA (C-89350)
Photograph of Inuit woman in traditional dress, taken by Geraldine Moodie, wife of RNWMP Superintendent J.D. Moodie

Sovereignty in Canada's northern reaches was yet far from secure, especially from covetous American eyes. A Boston columnist wrote in 1904:

> *Canada has, until now, failed to assert her sovereignty in an efficient manner, the American whalemen having prosecuted their industry there without interruption for over 70 years, so that they consider themselves entitled to continue fishing there in spite of Canada's contention of the contrary.*[8]

Apparently not all the American whalers were keen to submit to Canadian

SOURCE: NATIONAL ARCHIVES OF CANADA (C-1155)
Photograph of Inuit women at Fullerton Harbour 1904, taken by Geraldine Moodie, wife of RNWMP Superintendent J.D. Moodie (note the incorrect spelling of "belles").

SOURCE: NATIONAL ARCHIVES OF CANADA (C-1516)
Inuit of Fullerton Harbour in huge igloo, 1904, taken by Geraldine Moodie, wife of RNWMP Superintendent J.D. Moodie.

SOURCE: NATIONAL ARCHIVES OF CANADA (C-1198)
*Captain Bernier and crew at Winter Harbour, July 1, 1909.
Note the young muskox nuzzling Bernier's hand.*

rules, customs duties and taxation. Nor were the Norwegians to be ignored. Actively showing Canada's flag in the north was obviously more vital than ever to ensure her continued ownership of the Archipelago.

Bernier and Amundsen became aware of each other's presence in the north that winter. While the Norwegian Amundsen was over-wintering at Gjøa Haven, he sent an Inuk courier sledding over the tundra to Fullerton Harbour with mail for his homeland, having been informed by far-ranging Inuit that two big ships were over-wintering there. The return mail packet to Amundsen contained genial letters addressed to him from Superintendent Moodie, Captain Bernier and Captain Comer as well as gifts of 10 Eskimo dogs. Bernier gladly sailed Amundsen's mail out with him on the *Arctic* when he finally returned south. Although relations between Canadians and Norwegians on northern matters at the political level were somewhat strained, on the personal level they were quite cordial.

In July, Bernier and his men blasted a channel through the thick, icy mantle still blanketing Fullerton Harbour, releasing the *Arctic* from her icy prison into the open waters of the inland sea that is Hudson Bay, and he headed for home. After dropping anchor in several places on the way, Bernier docked once more in Quebec City on October 7, 1905, after an absence of 55 weeks.

> **The Bylot Island Proclamation**
>
> *This island, Bylot island, was graciously given to the Dominion of Canada, by the Imperial Government in the year 1880, and being ordered to take possession of it in the name of CANADA KNOW ALL MEN that on this day the Canadian Government Steamer "Arctic," anchored here, and I planted the Canadian flag and took possession of Bylot island in the name of Canada. We built a cairn to commemorate and locate this point, which we named Canada Point, after, and in honor of the first steamer belonging to the Canadian Navy.*
>
> *Being foggy no latitude was obtained. On the chart this point is located in Long. 80.50 west and 73.22 north latitude.*
>
> *From here the "Arctic" will proceed onward through Navy Board inlet, to the westward into Admiralty inlet, and from there westward to Port Leopold, where we will leave a record of our future work.*
>
> *Witnessed thereof under my hand this 21st day of August, 1906 A.D., in the fifth year of the reign of His Most Gracious Majesty King Edward VII.*
>
>
>
> *(signed)*
> *J.E. Bernier*
> *commanding officer*
> *by Royal commission*
> *F. Vanasse Historiographer*
> *Joseph Raoul Pepin M.D.*
> *Jas. Duncan Customs officer*
> *W.H. Weeks Purser*
> *Geo. R. Lancefield Photographer*[9]
>
>

In July of the following year, 1906, Bernier sailed again, this time on the first of three official two-year journeys aboard the *Arctic* in the name of the Dominion of Canada, specifically to establish and protect Canada's claim on the north. This time, Bernier was not only captain of the ship, he was also commander of the expedition, a much more prestigious and satisfying position for him than working under the orders of Superintendent Moodie. He sailed on these trips in 1906–07, 1908–09 and 1910–11, returning to southern Canada in between. Bernier inspected whaling stations, gave out licenses to whaling ships, collected duties for goods taken out of Canadian territory, and claimed a number of islands for the Dominion. He did this by anchoring off the shore of each of these islands and leading a landing party with several members of his crew as witnesses. He raised high the Canadian flag, read out a proclamation and left a copy of it, signed by himself and the witnesses, in a weather-tight container buried under a cairn of stones high enough to attract the attention of passing ships. The ship's photographer, George Lancefield, took the essential official photographs of the ceremonies as final proof of the acts. All the proclamations were virtually the same except for the particulars of the island being claimed (see that for Bylot Island on p. 110).

The 1906–07 trip was the first time Bernier had entered the High Arctic. By the time he left it in the fall of 1907, he had landed on the following islands and officially claimed them all for Canada: Bylot, Prince Patrick, Cornwallis, Lowther, Byam, Martin, Young, Melville, Davy, Eglinton, Garrett, Bathurst, Griffiths, Russell, Baffin and Ellesmere.*

In late October 1907, after over-wintering in Albert Harbour on Baffin Island near Pond Inlet, Bernier was back in Quebec City, drafting his trip report and making preparations for his next Arctic mission the following summer.

In 1908, Bernier again steamed into northern waters as captain of the *Arctic* and raised the Canadian flag on another group of islands including Banks, Victoria and King William islands before going into winter quarters at Winter Harbour on Melville Island. The year 1909 saw Bernier's master stroke in claiming Arctic islands for Canada—he invoked the sector principle.

In 1907, Canadian Senator Pascal Poirier had thought up the sector principle, which he invoked to justify Canada's claim over all the islands in the Arctic archipelago, whether visited by Canadians or not. According to Poirier's principle, Canada could claim title over a pie-shaped wedge north of her mainland between the 60th and the 141st longitudes and extending to the North Pole. There was, however, a small complication. The pie was not in keeping with geographic reality—it included a part of Greenland, but diplomats could work out a solution to that.

---

* The claiming of Ellesmere by Bernier was not really necessary since it had already been claimed by Low during his 1903–04 trip.

SOURCE: NATIONAL ARCHIVES OF CANADA (PA-118126)
Captain Joseph E. Bernier on his 1923 expedition.

At Melville Island's Winter Harbour, the sky was clear and crisp on July 1, 1909, a date whose significance was not lost on the captain of the *Arctic*. Bernier took his lead from Senator Poirier's sector principle and officially, with due pomp and circumstance, claimed for the Dominion, in the name of Edward VII, all the islands north of the Canadian mainland. From Bernier's report:

> *The first of July, Dominion Day, was celebrated by all on board; our flags were flying and the day itself was all that could be desired. At dinner we drank a toast to the Dominion and to the Prime Minister of Canada; then all assembled around Parry's rock to witness the unveiling of a tablet commemorating for all future explorers who venture into these remote regions the annexation by Canada of the whole of the Arctic Archipelago. I briefly referred to the important event in connection to the granting to Canada by the Imperial government on September 1, 1880, of all the British territory in the northern waters of the continent of America and the Arctic Ocean, from 60 degrees west longitude to 141 degrees west longitude, and as far north as 90 degrees, that is to say, the North Pole. I told my audience and traveling companions that we had annexed a number of islands, one by one, as well as vast stretches of continental territory, and that now we were about to establish claim to all the islands and territory within the coordinates I had just mentioned, and would henceforth be under Canadian jurisdiction. Three cheers would be given in honour of the Prime Minister, and the minister of Marine Fisheries of Canada, and the men dispersed for the balance of the day to enjoy themselves.*[10]

The marble tablet that Bernier left at Winter Harbour stated the following:

> *This memorial is erected today to commemorate taking possession for the Dominion of Canada, of the whole Arctic Archipelago, lying to the north of America, from long. 60W to 141W, up to lat. 90 North. Winter Hrb. Melville Island, C.G.S. "Arctic." July 1st 1909, J.E. Bernier, Commander, J.V. Koenic, Sculptor.*[11]

From this point on, Captain Bernier no longer considered it necessary to fly the flag on individual islands. Invoking the sector principle in 1909 took care of that, he reasoned. During his last official Arctic trip in 1910–11, during which he wintered over in Arctic Bay on Baffin Island, Bernier claimed no new land, but discharged his duties as customs collector, whaling license issuer and whaling station inspector, letting the American whaling captains know that they were plying their trade in Canadian waters.

Bernier left the service of the government in 1911, when the Canadian Arctic Patrols were discontinued, but that did not end his personal efforts to protect Canadian sovereignty in the north. Actually, it was more a matter of Bernier's interests and the Canadian government's interests coinciding. Undoubtedly, Bernier had been smitten by the North. When his government service ended, he embarked on a series of three private expeditions into the Arctic (1912–13, 1914–15, and 1916–17) and, among other things, set up trading posts on Baffin Island near Pond Inlet. This served his purpose—he was making his living in an area of the country that he loved. Canada's purpose was also served—a Canadian was actually established and carrying on a business in the Arctic. It was not a big contribution to the cause of sovereignty, but then, sovereignty rested on pretty shaky grounds at that time and every little bit counted.

Bernier was not yet finished with the Arctic, and it was fitting that the *Arctic* would once again be under his captaincy (though not his command) on his final trips north. The Canadian government was still not satisfied that its claim to sovereignty in the Arctic was as solid as it could be. It continued building its case on the international scene, resuming the Canadian Arctic Patrols (as Low's and Bernier's expeditions from 1903 to 1911 had become known). From 1922 to 1925, Bernier once again patrolled the Arctic to help establish Canadian police posts and to "show the flag." After Bernier's last Arctic patrol in 1925, they were continued for several decades under other captains, becoming known as the Eastern Arctic Patrol from 1927 on.

– 11 –
THE CANADIAN ARCTIC EXPEDITION, 1913–1918

VILHJALMUR STEFANSSON (1879–1962) WAS BORN in Manitoba in 1879, of Icelandic parents. He was a controversial figure. At one time, he commanded great respect as an extremely capable Arctic explorer, only to be later ignored and shunned by Canada's officialdom for gaffes partly due to his political naïveté, but also to the machinations of his enemies. He was far from naive while in his true element, traveling across the Arctic ice like an Inuk. He successfully adopted the Inuit ways of traveling, eating and dressing that made him self-sufficient as he ranged far into the unknown crannies of the Arctic archipelago. He once disappeared into the frozen mists for five months while the world gave him up for dead. When he finally emerged, he was quite surprised at the concern he had caused, for he and his companions had obviously eaten quite well, not having missed a meal during the whole five months as they lived off the land.

While exploring the Arctic coast of Canada on an early expedition, Stefansson's party came upon the beached carcass of a whale from which they cut out thick steaks for supper. Returning in the opposite direction over the same route the following year, they found that the whale was still there and again made a meal from the same carcass. The Arctic cold does not readily support bacteria and Stefansson knew that the meat when well cooked would not harm them. The taste, though, must have been another matter. He has been called the last of the old-time explorers.

In 1913, Stefansson led the Canadian Arctic Expedition (CAE) into the north for five years of exploration and new discoveries, but also to disaster:

> *The National Geographic Society had originally planned to finance our expedition, and it was only at the urgent request of the Canadian premier, the Right Hon. R.L. Borden, that the Society relinquished its direction of the enterprise. The Canadian Government felt that since the country to be explored was Canadian territory, it was only fitting that the expedition fly its flag and be financed from its treasury.*[12]

The CAE was really two separate expeditions, the northern one led by Stefansson himself, which was to explore unknown High Arctic regions, and the southern one under the leadership of Dr. R.M. Anderson, which was to gather more information about the already discovered Arctic coast of mainland Canada. It is the northern expedition that is of interest here.

On July 17, 1913, Stefansson sailed north from Esquimault, BC, in the former whaler, *Karluk*, with Captain Robert Bartlett at the helm, the same man who had been first mate on Peary's *Windward* when Sverdrup met the American explorer on Ellesmere Island in 1899. It was also Bartlett who sailed Peary into the Arctic on his much-disputed discovery of the North Pole in 1909.

Stefansson's plans quickly went awry. On August 13, 1913, the *Karluk* was caught by the ice near Herschel Island off the north coast of mainland Canada and was frozen into its Arctic embrace. Northern progress was finished for the year. This put a kink in Stefansson's plans, but a far greater kink was developing.

For the next month, the *Karluk* stayed pretty well in the same spot. Feeling that his ship was securely iced-in for the winter, Stefansson left her on September 20 with a party of men to go hunting caribou on the mainland. They were to be gone for 10 days. Stefansson never saw his ship again. On September 23, the *Karluk* in her prison of ice started drifting steadily westward toward Siberia until, on January 11, 1914, she was crushed by the relentless pressure of the ice and sank to the bottom. The lugubrious strains of Chopin's Funeral March played for the last time on Captain Bartlett's wind-up gramophone as the ship went down. Her crew escaped to the ice. What followed is a complete story in itself, dramatically described in Captain Robert Bartlett's book *The Last Voyage of the Karluk* and in survivor William MacKinlay's *Karluk*.

Survivors of the disaster made their way to Wrangel Island off the coast of Siberia. From there Captain Bartlett and an Inuk companion sledded 68 kilometres (42 miles) across the sea ice to Siberia and then along the coast of the Russian mainland, across Bering Strait to the vicinity of Nome, Alaska, from where news of the fate of the *Karluk* and its crew flashed out to the world over telegraph lines.

On September 8, 1914, almost six months after Bartlett and his Inuk companion had left Wrangel Island to seek help, the remaining survivors of the disaster were rescued by the ship *King and Winge* sent out from Alaska. They were then transferred to the *Bear*, the same ship that had picked up the pathetic survivors of Greeley's expedition off the coast of Ellesmere Island 30 years before.

SOURCE: NATIONAL ARCHIVES OF CANADA (C-86408)
The last photo of the doomed Karluk, *before she was crushed and sunk by ice, taken by F.W. Maurer.*

Of the 25 who were on the *Karluk* when it started its fatal drift, 14 survived; two parties of four men both disappeared shortly after the sinking, trying to make their way from the site of the wreck to Wrangel Island. Three died on the island while awaiting rescue.

Stefansson knew nothing of the fate of his ship. All he knew was that she had drifted away. He changed his plans accordingly and together with two companions spent the next four years exploring and discovering unknown land in the High Arctic, traveling, living and eating as the Inuit did while, unbeknownst to him until August 1915, World War I was raging in Europe. Reports had it that the war would soon be over, so Stefansson paid little heed to it.

In 1915, Stefansson's expedition discovered new land in the form of three islands until that time unknown to the world. These islands are known today by the names of Brock, after the chief of the Canadian

SOURCE: NATIONAL ARCHIVES OF CANADA (C-86406)
Photo of Vilhjalmur Stefansson, in 1914, taken by F. Johanssen.

Geological survey at the time, Borden and Mackenzie King,* after two Canadian prime ministers. In 1916, Stefansson again penetrated far into the north and discovered more land never before visited by white men—Meighen and Lougheed Islands. These five new islands were of course covered by the sector principle and thus previously claimed in 1909 by Bernier, but for a Canadian to actually discover them and claim them individually did not hurt either, just in case the sector principle was not universally recognized, which in fact it was not.

One of Stefansson's key skills in exploring the north was his ability to live off the land in the same way the Inuit did, hunting, eating game, wearing clothes made from caribou and seal skins, and sleeping in snow houses. The main advantage of this skill was that it permitted him a far greater range of operations than most explorers because he rarely ran out of supplies. This enabled him to disappear for many months at a time, only to reappear when least expected. More than once the press reported that he had been "saved" by the first people he met after a prolonged absence, when in fact he often put on weight on his trips.

In 1919 there were no Canadian patrols in the Arctic. Coincidentally or not, Knud Rasmussen, the famed half-Inuk explorer from Greenland, crossed over to Ellesmere Island that year and hunted down a large number of muskoxen. Canada protested, but Rasmussen's answer was that he considered Ellesmere a no-man's-land. The Danish government supported him. Canada's sector principle was not recognized by Denmark, among others, nor was it enforceable.

---

* At the time of the discovery in 1915, Stefansson thought he had discovered only two islands. An RCAF aerial survey in 1947 revealed that Brock Island was in reality two islands separated by a narrow strait. The second island was named after Canada's prime minister at the time of the survey.

– 12 –

THE TWENTY YEARS OF THE *ST. ROCH*, 1928–1948

Henry Asbjorn Larsen (1899–1964) was the next Canadian to be instrumental in weaving a stronger Canadian mantle of sovereignty in the north. It is ironic that he should have been born in Norway, only to become a protector of Canadian sovereignty in an area that included land first discovered, and still claimed at the time, by citizens of his country of origin.

Larsen came to Canada in 1923, having spent his youth at sea, including a trip into the Arctic on a commercial ship carrying trading goods. In 1927 he became a naturalized Canadian and the following year joined the RCMP, with a very specific idea in mind. He was aware that the RCMP was having a special patrol ship built for Arctic duty, the *St. Roch*. Until then, the RCMP Arctic patrol shipped out on commercial vessels and the Force felt it was about time it had its own ship. Larsen made up his mind that he would be part of the crew. His RCMP superiors were impressed by his seagoing experience and agreed with him. On June 28, 1928, three months after Larsen joined the RCMP, he sailed on the *St. Roch* out of Vancouver harbour on her maiden voyage, bound for the Arctic. The ship left harbour under a temporary skipper until she reached Arctic waters, after which a permanent captain was to be named. On August 28, two months after putting out to sea, the advantage of Larsen's seagoing experience was so apparent that he was made captain of the ship. Thus started Larsen's 20-year intimate association with the valiant *St. Roch*.*

---

* The *St. Roch* is preserved in the Maritime Museum in Vancouver, BC.

SOURCE: ROYAL CANADIAN NAVY, NATIONAL ARCHIVES OF CANADA (PA-121409)
The St. Roch *on patrol.*

The principle of the round bottom which the *Fram* had pioneered became recognized as essential if ships were to survive Arctic service. Like the *Fram*, the *St. Roch* was built with a salad bowl–shaped hull, which was essential for surviving the ice pack, but in a rough sea, she rolled and danced like a drunken sailor, soon separating the seamen from the landlubbers. She was only 32 metres (105 feet) long, sheathed with Douglas fir over a heavy frame and covered on top of that with a layer of Australian gum wood (iron bark, by its popular name). She was schooner-rigged, carrying a jib, foresail and mainsail. A 150-horsepower diesel provided the power needed to bull her way through the ice. The *St. Roch* had accommodations for 13, with some men sharing bunks when that number was exceeded, one man on duty, one in bed, and vice-versa when the watch changed.

For the next 20 years, from 1928 to 1948, there was not a year that Larsen and his ship were not in the Arctic. He made nine voyages during this period. On two of them, he did not overwinter in the Arctic, but on the seven others he and his crew spent anywhere from one to four winters solidly frozen into some protecting harbour along a barren, windswept Arctic coast. The RCMP was the official Canadian presence in the north at the time and as such had many and varied duties to perform, frozen in or not. In census years, Larsen and his men counted the Inuit and whites of the area they happened to be in. Any crimes, and particularly murders, had to be investigated and the guilty parties taken aboard to be brought to justice at the nearest court of law. Sick people were taken to hospital. Children were taken to school in far-away communities. Whaling and trading, especially by foreigners, had to be controlled. RCMP detachments had to be established.

In performing these duties, the RCMP effectively established and reinforced the sovereignty of Canada in these remote Arctic

SOURCE: CANADIAN PARKS SERVICE/
ENVIRONMENT CANADA
Henry Larsen during one of his many Arctic expeditions.

lands where no other government agency existed to do so. When the *St. Roch* was frozen in, RCMP members traveled long distances by dog-team to perform their duties. On February 13, 1941, a member of the crew, Frenchy Chartrand, suddenly collapsed and died of a heart attack. Larsen and his men were stunned. Frenchy was the only Roman Catholic member of the crew and his shipmates wanted him to benefit from the last rites of his church. The problem was that the nearest Catholic priest was Father Henry at Pelly Bay, some 640 kilometres (400 miles) away. On February 24, Larsen, accompanied by Constable Hunt and an Inuk, Ikualaaq, left by dog-team to fetch Father Henry. Larsen took the opportunity on his way to Pelly Bay and back to conduct a census of all the communities visited.

As soon as Father Henry was notified of Chartrand's death, he began making preparations for his long dog-sled trip to conduct a burial service back at the *St. Roch*. Larsen and his two companions preceded Father Henry, arriving back at the *St. Roch* on May 6 after 71 days' absence, having covered 1,900 kilometres (1,200 miles). Father Henry arrived almost two weeks later and conducted a Requiem Mass for Chartrand. Then the little procession moved onto the hill where Chartrand had been buried and Father Henry blessed his grave. A few days later, the priest set out for home.

Two of Larsen's northern trips in the *St. Roch* are particularly noteworthy. It was 1940, and World War II was raging in Europe. In the summer of that year Larsen departed Vancouver harbour with top-secret orders to sail toward Greenland via the Northwest Passage. Even his men did not know their destination until they were well out at sea. Denmark was occupied by the Germans and there were Allied plans afoot to protect the cryolite mine in Danish Greenland. The mineral was essential to the Canadian war effort for making aluminum. It appears that Larsen and the *St. Roch* would play some unspecified support role in these plans. Whatever that role may have been, the plan was not implemented and Larsen and the *St. Roch* bypassed Greenland and headed for Halifax instead. Twenty-seven months after leaving Vancouver harbour, he sailed into Halifax harbour, becoming the first captain ever to sail a ship through the Northwest Passage in the west to east direction.* It was not his fault that it took so long. He had been assigned some police duties during the crossing that had delayed him a year.

More than once Larsen and the *St. Roch* courted disaster in the uncompromising Arctic waters, but each time men and ship overcame the worst that the northern gods could throw at them. A "very close shave" is what Larsen dubbed an encounter with ice off the coast of Boothia Peninsula in the late summer of 1941:

> As the ice started to surround us, I back-tracked a bit and proceeded to the Boothia shore, as close as I dared, and then anchored by a small rocky islet, not much longer than the ship. We had barely settled down when a strong snowstorm came up from the northeast. Both anchors were let out and we prayed that they would hold. We had the engine going most of the night, with huge ice-flows crashing down on us. With the engine we managed to turn some of the floes aside, but we were in constant danger, and all of us spent the whole night on the fo'c'sle peering into the

---

* Norwegian Roald Amundsen had been the first to sail the passage in his tiny ship the *Gjøa*, in the east to west direction on a voyage lasting from 1903 to 1906.

darkness and the blinding snowstorm. This was, without doubt, one of the most difficult nights I, or any of the others, had experienced. The next morning the wind had changed to south and began driving the ice northward and us with it.

It was not long before we were completely locked in and drifting with the ice, dragging our anchors along.... Starting the engine, we tried to hold ground, but again got carried along helplessly by the tight-packed ice which now was held together by thick slush and snow. It was almost impossible to see or even to keep one's eyes open. It looked as if the elements were bent on our destruction.

Around four o'clock in the morning of September 6 the ice had carried us toward one of the big shoals we had spotted previously. The ship struck, pivoted twice and then remained on the shoal for a few minutes before she started to list to port. It looked as if she was going to topple completely when the ice started to climb right over the starboard side, now high out of water. Our port rail was already buried under the ice. This was a most uncomfortable situation, and we were all on deck trying to hang onto anything we could get hold of. All the time the deck seemed to be in instant danger of being completely buried.

I wondered if we had come this far only to be crushed like a nut on a shoal and then buried by the ice. Then suddenly a larger ice-floe came crashing through the darkness and hit the side of the ship, making it list even more. She was practically on her beam ends [on her side] *and it was our luck that the pressure did not let up just then, but kept on as if by a miracle until the ship was pushed over* [the shoal] *and a few moments later floated on an even keel in deep water.*[13]

Chances of survival or rescue were essentially nil if the *St. Roch* had foundered.

In July 1944, the *St. Roch* stood out from Halifax with Larsen at the helm to sail back through the Passage from east to west, arriving in Vancouver 86 days later. Larsen thus became the first man to sail the Northwest Passage in both directions and the first man to sail it in one season.

In the summer of 1947, Larsen and the *St. Roch* departed Vancouver on their last trip as protectors of Canadian sovereignty in the Arctic. They

SOURCE: NATIONAL ARCHIVES OF CANADA (C-70771)
Henry Larsen conning the St. Roch *through the ice.*

returned to port in October of that year, writing "finis" to a remarkable chapter of Canadian Arctic exploration. After this trip, Henry Larsen was through sailing Arctic waters in the *St. Roch*, but he was not yet through with the Arctic. In 1949 he was made Officer Commanding "G" Division, the RCMP division that controlled the police work in all of northern Canada. He was headquartered in Ottawa. For the first time in his life, Hanorie Umiarjuaq—Henry with the Big Ship—as the Inuit knew him, sat at a desk in an office.

– 13 –

CANADA AND NORWAY NEGOTIATE

IN 1926, THE CANADIAN GOVERNMENT BECAME officially concerned about the status of the so-called Sverdrup Islands including Axel Heiberg, Ellef Ringnes and Amund Ringnes, as well as that part of Ellesmere Island that Sverdrup had discovered during his 1898–1902 expedition. In 1926, Canada sent a diplomatic note to Norway, inquiring about the basis that country might have to claim the territory in question. The Norwegians took their time in answering. In 1928, Norway responded via letter from the Norwegian consul in Montreal reserving all rights of Norway under international law over the Sverdrup Islands. However, after further diplomatic negotiations, on August 8, 1930, the Norwegian Chargé d'Affaires in London sent the following note to the British Secretary of State for Foreign Affairs:

> Sir—Acting on instructions from my Government I have the honour to request you to be good enough to inform His Majesty's Government in Canada that the Norwegian Government, who do not as far as they are concerned claim sovereignty over the Sverdrup Islands, formally recognises the sovereignty of His Britannic Majesty over these Islands.
>
> At the same time my Government is anxious to emphasize that their recognizance of the sovereignty of His Britannic Majesty over these islands is in no way based on any sanction whatever of what is named the "sector principle."[14]

It is not known for certain what motivated the Norwegian government to renounce ownership of the Sverdrup Islands, but it is likely Norway thought that defending the islands against Canadian claims was neither practical nor economically feasible. They were simply too far away and not worth the trouble. Whatever the reason, the decision to give up the islands was a bitter disappointment to Otto Sverdrup, who desperately wanted his country to keep them.

It is interesting to note that although Norway recognized British sovereignty over the islands, it did not recognize that it was by virtue of the sector principle, which Norway repudiated. The sector principle might have served the purposes of Canada, but she was alone in this, with the exception of perhaps the Soviet Union, which found some value in it for claiming some islands north of Siberia. However, the two most important players in any possible sovereignty dispute with Canada, Norway and the United States, have never recognized the sector principle. They have both, in fact, expressed disapproval of it. Nor has the validity of the principle ever been tested in a court of international law, and indeed, it appears to rest on somewhat shaky ground.

Donat Pharand, professor of International Law at the University of Ottawa, had this to say about the sector principle in 1981: "However, the general opinion is that the sector theory has no legal validity as a source of title in international law and cannot serve as a legal basis for the acquisition of sovereignty over land and, *a fortiori*, over sea areas."[15] All the more reason to ensure sovereignty by other means.

There still existed, however, the somewhat irregular situation that a Norwegian had done a major job of discovering, surveying and mapping land that was now Canadian. The value of Sverdrup's work was officially recognized by the Canadian government:

> *During the past Session of the Canadian Parliament Supplementary estimates were approved covering the grant of $67,000 (£31,700) to Commander Otto Sverdrup, the famous Norwegian explorer, in recognition of his explorations and discoveries in the Canadian Arctic Archipelago and for the purchase of the original maps, notes, diaries and other documents relative to his explorations and discoveries in that area. The grant has been made by the Canadian Government in accordance with this provision.*

> *Commander Sverdrup's survey of the Axel Heiberg, Amund Ringnes, Ellef Ringnes and King Christian Islands, which he has carefully mapped, together with the scientific research carried out there, represent a remarkable contribution to science which has led to important results in many different branches of knowledge. They have been most useful to the Department of the Interior and to the Royal Canadian Mounted Police in connection with their annual patrols of northern Canada. The Canadian Government therefore decided to make ex gratia a pecuniary grant to Commander Sverdrup, which might at the same time be accepted as a consideration for the delivery of all original maps, records, diaries, and other material in the explorer's possession."*[16]

Although the official record indicates that the grant to Sverdrup was for the delivery of his original maps, records and diaries, several pieces of official correspondence show that Canada was in fact purchasing much more than mere pieces of paper. A letter dated October 17, 1930, from O.S. Finnie of Canada's Department of the Interior to O.D. Skelton, Undersecretary of State for External Affairs, Ottawa, leaves absolutely no doubt as to the real motive behind the $67,000 grant. It was to maintain a continuous, unbroken thread of sovereignty from 1880, when the British Crown ceded the Arctic to Canada, right through to 1930 and hopefully forever after. The continuity of the thread gave it its strength and Canada did everything in its power to demonstrate that the thread had never been broken:

> *The main objective of entering into our negotiations with Sverdrup was for the purpose of securing from the Norwegian Government a recognition of the British Sovereignty in that portion of the Arctic north of the North American continent.*[17]

But O.D. Skelton did not need Finnie to tell him this. On September 23, 1930, Skelton himself had written the following to the British representative in Oslo, the Hon. C.J.F.R. Wingfield, C.M.G., H.B.M. Envoy Extraordinary and Minister Plenipotentiary:

> *In view of the probable early conclusion of the negotiations, we desire to arrange for the payment as early as possible to Commander Sverdrup of the grant of $67,000 made by the Canadian Parliament, **conditionally on the reaching of a satisfactory agreement as to the title to the islands**.*[18]
> (emphasis added)

Nothing could be clearer. In spite of the officially declared position that Canada was buying the originals of Sverdrup's maps and records with the grant, it is evident that Canada was really ensuring a recognition of clear and continuous title to the Arctic islands. It became even more evident when Canada returned Sverdrup's documents to his widow upon her request after World War II even though obtaining them had been the stated reason for paying the $67,000.

In November 1930, Canada wrote private citizen Sverdrup a cheque for $67,000 for his work and the case was closed. The islands were officially Canada's at last. By 1930, the thread of Canadian sovereignty over the Arctic Islands that had started in 1880 was still unbroken.

Two weeks later Otto Sverdrup lay on his death bed. He died on November 26, 1930, his widow adequately, if not luxuriously, provided for by Canada's grant.

Norway and Canada can truly be proud of the role played by Norwegian and Canadian heroes, and their ships, in the discovery, exploration and protection of the Canadian Arctic islands—Sverdrup in the *Fram*, Low in the *Neptune*, Bernier in the *Arctic*, Amundsen in the *Gjøa*, Stefansson in the doomed *Karluk*, Larsen in the *St. Roch:* these names in particular stand out among the many who were called into the Arctic at the end of an era of iron men and wooden ships.

Epilogue

Otto Sverdrup of Norway visited the Canadian High Arctic in the years 1898–1902, exploring and claiming several of its islands, but he was not the first nor the last to disregard Canadian sovereignty in the Arctic.

Following Sverdrup's triumphant return to Norway from his voyage of discovery in the Arctic, another Norwegian, Roald Amundsen, from 1903 to 1906, navigated the Northwest passage through Canada, overwintering twice in Gjøa Haven on King William Island.

Before and after Sverdrup, from 1890 to 1910, American whalers used Herschel Island off the Yukon coast for overwintering between whaling seasons. They also overwintered in Hudson Bay.

In 1919, Knud Rasmussen a half-Inuk explorer from Greenland crossed over to Ellesmere Island and hunted down a large number of musk oxen. Canada protested, but Rasmussen's answer was that he considered Ellesmere a no-man's-land. The Danish government supported him.

In 1969 the USS *Manhattan* sailed through the Northwest Passage to test the feasibility of oil tankers using the route to attain the far east. In doing so, it came within five kilometres of Canadian land in the narrowness of Prince of Wales Strait between Banks and Victoria islands. The American ship did not ask permission to transit through the Northwest Passage. However, to avoid creating a precedent, Canada diplomatically gave the ship permission, without being asked.

Inuit in Cambridge Bay, Resolute and other locations have seen, and in some cases photographed, American and Russian submarines navigating channels between Arctic islands.

These are only a few of the more than 25 instances of violation of Canadian sovereignty that have occurred since the Arctic was turned over to Canada by Great Britain in 1880. Even today, Denmark has not recognized that half of Hans Island in the strait between Ellesmere Island and Greenland belongs to Canada as has been determined by an international survey. It is not a large island, but the question is an important one of oil-drilling rights.

From 1880, when Canada was given the High Arctic for safekeeping, until 1902 when Sverdrup's claims deeply shocked Prime Minister Wilfrid Laurier and his government into action, Canada had fallen into a state of complacency about its northern possessions. It learned to its chagrin that complacency is a dangerous state. In 1930 Canada got off easy—a mere $67,000 settled the issue after years of diplomatic haggling. It is certain that we would not be that fortunate the next time.

In 1903 Canada learned that legal arguments and historic paper claims are no competition against effective occupation when it failed in its own claims to the Alaskan "panhandle" that hangs down near the west coast of Canada. History has a habit of repeating itself. Has Canada learned from such cases? Not enough. We are once more in a state of complacency about our northern reaches. Global warming is upon us and within fewer years than we can imagine, the fabulous Northwest Passage that so many died for in the nineteenth century may finally be a reality. There will be a plethora of interested potential users. Will Canada be in a position to protect its own interests? Only if our government takes the steps necessary to ensure that it will be, starting now.

We all know that popular maxim, "Use it or lose it." In its complacency this is what Canada has not yet learned. It would be dangerous not to pay attention to history. Effective occupation has to be demonstrated. Much has to be done to achieve effective use and occupation not only of the land, but also the waters. This will take serious government commitment and involvement.

I surely don't pretend to have answers to all that will have to be done, but I would at least suggest a measure that could be implemented now as a start and which would have a reasonable cost and a beneficial effect. There are many scientific matters that have yet to be adequately addressed in our northern lands. An increased investment in science may be the cheapest, and most effective, immediate means of establishing a sovereign base for our northern lands and seas. In-depth scientific research in the Arctic, coupled with the presence of the many people required for this research, would be a fine way of developing Canada's scientific cadre, including the many young students needed for this work, while increasing the interest and knowledge of all the people of Canada in our northern reaches. It would yield an excellent beginning to the demonstration of effective occupation as well as a sound foundation for future efforts.

It is crucial that Canada not be caught napping again regarding security of its sovereignty over Arctic lands, and particularly, waters.

George D. Hobson
Former Director of the Polar Continental Shelf Project

ENDNOTES

1. A.P. Low, *The Cruise of the Neptune* (Ottawa: Government Printing Bureau, 1906).
2. Ibid.
3. Owen Beattie, *Frozen in Time* (Saskatoon: Western Producer Prairie Books, 1988).
4. J.E. Bernier, *Master Mariner and Arctic Explorer* (Ottawa: Le Droit, 1939), 305–06.
5. Yolande Dorion-Robitaille, *Captain J.E. Bernier's Contribution to Canadian Sovereignty in the Arctic* (Ottawa: Ministry of Indian and Northern Affairs, Ottawa, 1978).
6. Ibid.
7. Ibid.
8. Ibid.
9. Ibid.
10. Ibid.
11. Ibid.
12. Bartlett, Robert, Ralph Hale. *The Last Voyage of the Karluk* (Boston: Small, Maynard and Company, 1916).
13. Henry A. Larsen, *The Big Ship* (Toronto: McLelland and Stewart Ltd., 1967).
14. National Archives of Canada, MG 30 B75, the Larsen Collection, Vol. 3, Larsen's file of trip reports, 1934–47.
15. M. Zaslow (ed.), *A Century of Canada's Arctic Islands, Canada's Jurisdiction in the Arctic* (Ottawa: Royal Society of Canada, 1981), 117.
16. National Archives of Canada, RG22, Vol 254, File 40-8-1 Pt4, Information Bulletin of the Office of the High Commissioner of Canada, London, January 7, 1931.
17. National Archives of Canada, RG22, Vol. 254, File 40-8-1, Pt4, Letter from O.S. Finnie to O.D. Skelton, October 17, 1930.
18. Ibid.

BIBLIOGRAPHY

Alcock, F.J. 1948. *A Century in the History of the Geological Survey of Canada*. Ottawa: King's Printer.

Bartlett, Robert and Ralph Hale. 1916. *The Last Voyage of the Karluk*. Boston: Small, Maynard and Company.

Bassett, John. 1980. *Henry Larsen*. Don Mills, ON: Fitzhenry and Whiteside Limited.

Bernier, J.E. (translated by Paul Terrien). 1983. *Les Memoires de J.E. Bernier*. Montreal: Les Quinze.

Dorion-Robitaille, Yolande. 1978. *Captain J.E. Bernier's Contribution to Canadian Sovereignty in the Arctic*. Ottawa: Ministry of Indian and Northern Affairs, Canada.

Fairley, T.C. 1959. Sverdrup's Arctic Adventures. London and Toronto: Longmans, Green and Co.

Fairley, T.C. and Charles E. Israel. 1957. *The True North*. London: MacMillan.

Francis, Daniel. 1986. *Discovery of the North*. Edmonton: Hurtig Publishers.

Guttridge, Leonard F. 2000. *Ghosts of Cape Sabine*. New York: G.P. Putnam's Sons.

Horwood, Harold. 1977. *Bartlett the Great Canadian Explorer*. Toronto: Doubleday Canada Limited.

Kobalenko, Jerry. 2002. *The Horizontal Everest*. Toronto: Penguin Books Canada Ltd.

Larsen, Henry A. (with Frank R. Sheer and Edvard Omholt-Jensen). 1967. *The Big Ship*. Toronto/Montreal: McClelland and Stewart Limited.

Low, Albert Peter. 1906. *The Cruise of the Neptune*. Ottawa: Government Printing Bureau.

MacDonald, R.St.J. (ed.). 1966. *The Arctic Frontier*. Toronto: University of Toronto Press.

McKinlay, William Laird. 1976. *Karluk*. London: Weidenfeld and Nicolson.

Robertson, Gordon. 1988. *The North and Canada's International Relations*. Ottawa: Canadian Arctic Resources Committee

Shackleton, Edward. 1937. *Arctic Journeys*. London: Hodder and Stoughton Limited.

Sverdrup, Otto. 1904. *New Land*, Vols. I and II. London: Longmans, Green, and Co.

Thompson, John B. 1974. *The More Northerly Route*. Ottawa: Indian and Northern Affairs—Parks Canada.

Zaslow, Morris (ed.). 1981. *A Century of Canada's Arctic Islands*. Ottawa: Royal Society of Canada.

INDEX

(bold page numbers indicate photographs)

A
Amund Ringnes Island, 125; explored, 47, 49, 54, 62, 70
Amundsen, Roald, 101, 109
Anderson, R.M., 116
Archer, Colin, 9
Arctic: 1906-1911 expeditions, 110–113; 1905 expedition, 105, 109; picture, **106**; picture of crew, **109**
Arctic Journeys (Shackleton), 45–46
Arctic night, 24–25, 26, 28
Axel Heiberg Island: ceded to Canada, 125; determining it's an island, 44, 54, 62, 67; explored, 47–50, 62; named, 29

B
Bartlett, John, 20n
Bartlett, Robert, 20n, 116–117
Bartlett, S.W., 95, 97
Baumann, Victor, 7, 36; camp duties, 25, 41; explores Beechey Island, 76, 78–80; searches for land route on Simmons Peninsula, 43, 50; sets up depots, 62, 76, 78; shooting excursion, 56, 57; visits Peary, 28–29
Bay, Edvard, 8, 25, 36, 85; crosses Ellesmere, 29; dredging excursion, 82–83, 84; guards bear meat, 37, 41–42, 53; meets Peary, 19–20; writes novel, 75–76
Bear, 117
Beechey Island, 76, 78–80, 100–101
Belcher, Sir Edward, 101
Bernier, Joseph Elzear, 101, 107; 1906-1911 expeditions, 110–113; background and character, 103–105; contact with Amundsen, 109; picture, **112**
Bignell, John, 93–94
Boothia Peninsula, 122–123
Borden, Dr., 98
Borden, Robert L., 116
Braskerud, Ove, 8, 25; crosses Ellesmere, 30, 32; death, 36–37; grave cross, **38**, 85; picture, **37**
Bylot Island Proclamation, 110

C
Canada, Government of: makes claim on north, 105, 110–111, 113, 116; negotiates with Norway, 125–128; reaction to Norwegian exploration, 91–92, 96; urged to act in north, 130
Canadian Arctic Expedition, 115–118
Canadian Arctic Patrols, 113
Century in the History of the Geological Survey of Canada, A (Alcock), 93–94
Chartrand, 'Frenchy,' 121–122
Choque, Charles, 13–14
Chronometers, 79–80
Coal, 63
Comer, George, 98, 107, 109
Cryolite mine, 122

D
Denmark, Government of, 118, 129
Dogs, Eskimo: character, 12–14, 24, 86; hunting polar bear, 39–41, 50; and wolf attack, 68–70
Ducks, eider, 71–72

E
Ellesmere Island: Canada takes possession of, 99–100; crossed, 29, 30, 32; explored, 43, 44–47
Era, 95, 96, 97, 98
Erik, 99
Eskimo dogs. *see* Dogs, Eskimo
Eskimos. *see* Inuit

F
Faribault, Dr., 98–99
Finnie, O.S., 127
Fosheim, Ivar, 8; camp duties, 25, 41, 58, 62, 70, 76; explores Axel Heiberg, 47–50; explores Baumann Fiord, 62–68; explores Beechey Island, 76, 78–80; explores Ellesmere, 43, 44–47; helps Olsen, 60, 61; trip through ice tunnel, 51–52
Fram: awning fire, 53–54; battles ice, 32–33, 55–57, 72–73, 82, 83–84, 86; description, 5, 9; pictures, **29, 74**; prepared for overwintering, 23–24
Franklin, Lady, 101

Franklin expedition, 78–79
Freuchen, Peter, 12–13

G
Gauss, 105. see also *Arctic*
Gjøa, 101
Godhavn, Greenland, 85–86
Great Britain, Government of, 125, 127
Greely, Adolphus, 26–28

H
Hall, Charles, 15
Hares, Arctic, 64–66
Hassel, Sverre, 8; camp duties, 62, 76; explores Amund Ringnes Island, 47, 49, 54, 62, 70; explores Ellesmere, 43, 44–47; picture, **55**; visits Peary, 28
Heiberg, Axel, 3
Hell Gate, 42–43
Hendriksen, Peder Leonard, 8, 25, 35, 37, 62; dredging excursion, 82–83, 84
Henry, Father, 121–122
Hunt, Constable, 121

I
Ice blink, 10
Ikualaaq (Inuk guide), 121
Inglefield, E.A., 34
Inuit, 85–86, 96–97, 107; pictures, **107, 108**
Inuit settlement, 71–72
Isachsen, Gunerius Ingvald, 8, 25, 62; crosses Ellesmere, 30, 32; dredging excursion, 82–83, 84; explores Amund Ringnes Island, 47, 49, 54, 62, 70; explores Ellesmere, 43, 44–47; mistakes Axel Heiburg as extension of Ellesmere, 54, 62; picture, **55**

J
Jackson, Frederick, 5
Jeannette, 4, 104
Johansen, Frederik, 5

K
Karluk, 116, **117**
King and Winge, 117
Kobalenko, Jerry, 32n

L
Lancefield, George, 111
Larsen, Henry Asbjorn, 11; patrols Arctic, 119–124; pictures, **121, 124**
Laurier government, 91–92, 96, 105

Lindström, Adolph Henrik, 8
Lockwood, J.B., 29n
Low, Albert Peter: background and character, 93–94; explores Arctic, 94, 95, 97–98, 99–100, 101; picture, **97**

M
MacKinlay, William, 116
Maps, 37–38
Mary, 79, 80
McClintock, Sir Francis, 101
Moodie, J.D., 107, 109; on *Arctic*, 101, 105; on *Neptune*, 95, 96, 99; report on Faribault's insanity, 98–99
Muskoxen hunting, 19, 21–23, 63–64, 81

N
Nansen, Fridtjof, 4–5, 7n, 12
Neptune, 94–99, **95**; pictures of crew, **96, 97**
Nödtvedt, Jacob, 8, 37; camp duties, 25, 70
North Pole exploration: Bernier's interest in, 104, 105; early Norwegian attempts, 3–5; Peary's attempts, 20–21
Northwest Passage, 96, 122, 123, 129, 130
Norway, Government of, 125–128
Norwegian Magnetic Pole Expedition, 101

O
O'Connell, James, 99
Olsen, Karl, 8, 68; camp duties, 41, 70, 86; dislocates arm, 59–61; looks for Greely's camp, 26, 28

P
Peary, Robert Edwin, 20–21, 28–29
Pharand, Donat, 126
Poirier, Pascal, 111
Polar bear hunting, 39–41, 50
Polar night, 24–25, 26, 28
Polaris, 15–16, 104
Proteus, 26

R
Raanes, Olaf, 7–8, 50; camp duties, 25, 70; explores Baumann Fiord, 62–68; explores Beechey Island, 76, 78–80; sets up depots, 62, 76, 78; shooting excursion, 56, 57

Rasmussen, Knud, 118
Ringnes, Amund, 3
Ringnes, Ellef, 3
Ross, John, 79
Royal Canadian Mounted Police (RCMP), 119, 121
Royal North West Mounted Police (RNWMP), 92

S
Schei, Per, 28, 39, 43, 62; background, 8, 87; explores Baumann Fiord, 62–68, 70; explores Beitstad, 30, 31; explores Eureka Sound and north, 76, 78, 80–82; hunts muskoxen, 22–23; maps Devon Island, 71–72; and wolves, 68, 69
Scotty (Inuk pilot), 97, 98
Sea ice, 45–47
Seals, 25–26
Sector principle, 111, 112–113, 118; questioned, 125, 126
Shackleton, Edward, 45–46
Simmons, Herman Georg, 8, 30, 53, 85; dredging excursion, 82–83, 84; helps Olsen, 59, 60; hunts muskoxen, 22–23
Skelton, O.D., 127
Smith-Johannsen, Herman, 9n
St. Roch, 119–124, **120**
Stefansson, Vilhjalmur: character, 115; explores high Arctic, 117–118; leaves the *Karluk*, 116; picture, **118**
Stolz, Rudolph, 8, 43, 70; maps Devon Island, 71–72
Svendsen, Johan, 8, 25; illness and death, 30–32; looks for Greely's camp, 26, 28
Sverdrup, Mrs., 128
Sverdrup, Otto: 1893-96 expedition, 3, 4–6; death, 128; efforts recognized by Canada, 126–127; enters Flagler Fiord, 30; and Eskimo dogs, 12, 86; explores Axel Heiburg, 47–50; explores Baumann Fiord, 62–68, 70; explores Ellesmere, 29, 33, 43, 44–47; explores Eureka Sound and north, 76, 78, 80–82; and Franklin boat, 76, 78–79, 80; on hunting muskoxen, 21–23; maps Devon Island, 71–72; meets Robert Peary, 19–20; navigates *Fram*, 11–12, 14–15, 16, 55–57, 72–73; and Olsen's injury, 59–61; picture, **4**; on sledding on thin ice, 46–47; and Svendsen's death, 31, 32; tribute to Braskerud, 37, 85; trip through ice tunnel, 51–52
Sverdrup Islands, 125–128. *see also* Axel Heiburg Island;Amund Ringnes Island
Sverdrup's Second Arctic Expedition (*see also Fram;* Sverdrup, Otto; and individual members of the expedition): achievements, 87–88; adjusts chronometers, 79–80; buries messages, 76, 77; celebrates special days, 26, 49–50, 68, 81; crew, 7–9; crew picture, **8**; establishes meat cache, 37–39; exploration, 26, 28, 30–32, 35–36, 43–50, 54, 62–68, 70–72, 76, 78–82; finds coal, 63; fire on *Fram*, 53–54; frozen in, 55–56, 73, 74, 84; homecoming, 86–87; makes supplies for winter, 25, 41, 59, 70; misses mail drops, 33, 81–82, 85; moves to avoid fifth year in ice, 76, 78–80, 82–83; sanding the ice, 76, 82; setting off, 7, 9–10, 11–12; winter anchor spots, 16, 34, 57

T
tree stumps, 63

U
USS Manhattan, 129

W
Walrus hunting, 17–19
Whaling, 92, 95, 96, 107, 109, 113, 121, 129
Windward, 20–21, 33, 101
Wingfield, C.J.F.R., 127
Wolves, 21, 69–70
Wood, 63

Y
Young, Allen, 79

About the Author

Gerard Kenney—his friends call him Gerry—was born in St. Rémi d'Amherst not far from Mont Tremblant, Quebec in 1931. Though a Canadian, he spent the first 16 years of his life in New York City, except for the months of July and August which he enjoyed in the small French-Canadian village of his birth. In 1948, he returned to his native Canada and has lived there ever since.

Gerry's work as a telecommunications engineer has taken him to many countries of the world as well as to the northern reaches of his native land. Working for Bell Canada in the 1960s and 1970s, Gerry was responsible for the engineering aspects of the telephone system based on short-wave radio that served the eastern half of the Northwest Territories, Labrador and Nouveau Quebec.

In the late 1960s while he was travelling on Ellesmere Island, an RCMP officer in Grise Fiord showed him the horizontal member of a wooden burial cross which had been found nearby. It was in memory of a Norwegian sailor, Ove Braskerud, who had left his bones in the frigid waters of nearby Harbour Fiord in 1899. Braskerud had been a member of the 1898–1902 Sverdrup expedition aboard the Norwegian ship *Fram* which discovered and explored the Arctic islands lying north of the Canadian mainland. That chance encounter with Braskerud's cross eventually led to this book.

This is Gerry's second book about the Arctic. His first, published in 1994, is entitled *Arctic Smoke & Mirrors*.

Gerry is now retired which gives him more time to pursue his interest in writing about, and exploring, things northern. In 1999, he and a friend canoed a part of the path of the fatal 1903 Hubbard and Wallace expedition in Labrador.

Gerry lives in Ottawa and has two daughters, Amanda in Montreal and Jessica, accompanied by young Cara, in Ottawa.